AF686651

Evaporation, Condensation and Heat transfer

Evaporation, Condensation and Heat transfer

Editor

Petros Antonis

scitus
academics

Evaporation, Condensation and Heat transfer
Edited by **Petros Antonis**

ISBN: 978-1-68117-308-5
Library of Congress Control Number: 2016937617

© 2017 by
SCITUS Academics LLC,
www.scitusacademics.com
Box No. 4766, 616 Corporate Way,
Suite 2, Valley Cottage,
NY 10989

This book contains information obtained from highly regarded resources. Copyright for individual articles remains with the authors as indicated. All chapters are distributed under the terms of the Creative Commons Attribution License, which permits unrestricted use, distribution, and reproduction in any medium, provided the original author and source are credited.

Notice

Reasonable efforts have been made to publish reliable data and views articulated in the chapters are those of the individual contributors, and not necessarily those of the editors or publishers. Editors or publishers are not responsible for the accuracy of the information in the published chapters or consequences of their use. The publisher believes no responsibility for any damage or grievance to the persons or property arising out of the use of any materials, instructions, methods or thoughts in the book. The editors and the publisher have attempted to trace the copyright holders of all material reproduced in this publication and apologize to copyright holders if permission has not been obtained. If any copyright holder has not been acknowledged, please write to us so we may rectify.

Printed in United States of America on Acid Free Paper

Preface

Heat is the vibrating particles ' film energy. If we heat the particles at one end of a material, the particles vibrate more at that end (have more kinetic energy) and bump into the adjacent particles. They collide with their neighbors so that the energy passes from one particle to another through the material. Evaporation and condensation are two processes that convert matter to state. Three different states of matter can be found: solid, liquid or gas. During evaporation, matter changes from liquid to gas and material changes from gas to liquid during condensation. All matter is made up of small moving particles called molecules. Evaporation and condensation occur when the molecules gain heat energy or lose it. Evaporation takes place when the liquid is heated. The heat gives the liquid molecules more energy. With this energy, the molecules move faster. If you gain enough energy, the molecules near the surface break away. They escape the fluid and as gas they enter the air. If molecules cool down in a gas, condensation occurs. Since molecules lose heat, power is lost. As a result, they slow down. They draw closer to other gas molecules. Finally, these molecules collect to form a liquid. The theoretical analysis and modeling of heat and mass transfer rates in evaporation and condensation processes are important issues in the design of a wide range of industrial processes and equipment.

The book highlights various aspects of the current problems of evaporation modeling, water vapor condensation, heat transfer and fluid flow. The approaches would also apply to various industrial applications. The innovative ideas and information described in this book will be helpful in finding a sustainable solution for readers in an industrialized society.

Table of Contents

1

CHAPTER

EVAPORATION INDUCED THERMAL PATTERNS IN FLUID LAYERS: A NUMERICAL STUDY

Claire Perfetti, Carlo Saverio Iorio

Service de Chimie-Physique E.P., Université Libre de Bruxelles, Bruxelles, Belgium

ABSTRACT

Liquid layers evaporating under the influence of a gas shear flow presents a non-uniform distribution of the evaporation rate all along the interface. Being the evaporation an endothermic process, a thermal gradient along the interface is generated and thermo-capillary flows are induced. Hence, two opposite mechanisms regulate the movement of the interface: the shear stress of the gas that entrains the interface in the direction of the flow and the thermo-capillary stress that forces the interface to move against the flow direction. The composition of these mechanisms at the interface generates an unstable thermal patterning. The dynamic evolution of the patterning and the relative evaporation rate are strongly influenced by the flow rate of inert gas, the layer thickness and the liquid thermo-physical properties. The goal of the present work is to study numerically how the evaporation process is influenced by the above-mentioned mechanisms. The focus will be on the evolution of the thermal patterning at the interface and the assessment of the main factors influencing the computed evaporation rate.

Keywords: Heat and Mass Transfer, Evaporation, Marangoni Convection, Thermo-Capillarity, Boundary Layers

1. INTRODUCTION

Thermal and/or solutal patterning of evaporating fluid layers is becoming an intriguing field of research for its promising applicability in many industrial applications, including the rapidly expanding domain of bio-technology [1] and tissue engineering [2] . Also, compact, enhanced heat exchangers for cooling of micro-electronics often take advantage of the intrinsic endothermic character of the evaporation process. Despite the relatively great number of theoretical models [3] -[5] and semi-empirical assessments, the evaluation of the actual heat and mass transfer rate for fluid layers evaporating in presence of a flow of inert gas is still an open domain of investigation. As a matter of fact, at the interface of the evaporating layers, the gas flow induces an additional shear stress that couples with the thermo-capillary one induced by the evaporation process. This coupling can alter significantly the thermo-fluid-dynamic field at the interface, hence change the heat and mass transfer rate. To the best of our knowledge, many studies have been devoted both to thermo-capillary convection [6] and evaporation of a volatile liquid in presence of a still inert gas [7] [8] , but only few contributions could be found when an additional shear flow is presented [9] -[12] . Also, the influence of the shear stress at the interface has been thoroughly studied in cavity flow problems, by neglecting phase change processes. The role that these three effects-evaporation, thermo-capillarity and shear-stress could have, when acting contemporarily, remains not completely understood [3] . In this paper, we will study numerically the case of a volatile fluid evaporating when subject to a shear-stress flow. Focus will be given to the role that the evolution of the concentration boundary layer could have in determining the effective heat and mass transfer rate at the interface.

Let consider a horizontal, isothermal layer of pure volatile liquid whose interface is exposed to a flow of inert gas. To simplify the modeling efforts, the interface is considered to remain completely flat and un-deformable while being subject to a tangential stress. If evaporation is neglected, the liquid layer will show a bulk movement induced by the shear stress acting on the interface because of the presence of the inert gas flow. A steady state is then reached according to the continuity of the shear stresses at the interface. When evaporation is took into account, and then the interface begins to cool. This cooling is not uniform all along the interface because the flow of gas removes the vapor more effectively at the inlet border than elsewhere along the interface. If we focus our attention on the liquid layer, we deal with a horizontal thermal gradient that acts along the interface in the direction of the flow and a vertical one due to the general cooling of the interface that makes the layer sensitive to convective instabilities. For many fluids at the surface tension a monotonic decreasing function of the temperature, the horizontal thermal gradient is responsible for a thermo-capillary driven global bulk motion of the liquid in the direction contrary to the inert gas flow. For the sake of simplicity, we neglected the influence of gravity. If present, when the vertical thermal gradient would have exceed a critical value, a bulk convective motion in the form of recirculation rolls should have also arisen. This convective motion has strong effects on the temperature distribution of the liquid bulk phase and, as a consequence, the heat flux exchanged between the liquid and the wall can significantly change respect to the pure conductive state.

2. NUMERICAL MODEL

Numerical simulations have been performed to assess the effects that the coupling between shear flows, evaporation and thermo-capillary convection could have in the heat and mass transfer characteristics at the interface of the working fluid. The commercial software Fluent complemented by the mesh generator Gambit has been used as the basic CFD solver. Specific custom-made C code, the so-called User Defined Functions, has been implemented to take into account for evaporation and thermo-capillary effects that are not included in the standard models [13] . Numerical simulations have been performed on the 3D geometry (Figure 1) consisting of a square liquid pool of 50 × 50 mm and a depth of 1.6 mm.

It is important also to stress that the two-layer problem has been studied. This means that also the thermo- fluid dynamic fields in the gas channel superposed to the liquid pool have been computed.

The number of computational points has been chosen according to a mesh sensitivity rule: tests with different mesh sizes have been performed. Solution was calculated on meshes of increasing computational points' number.

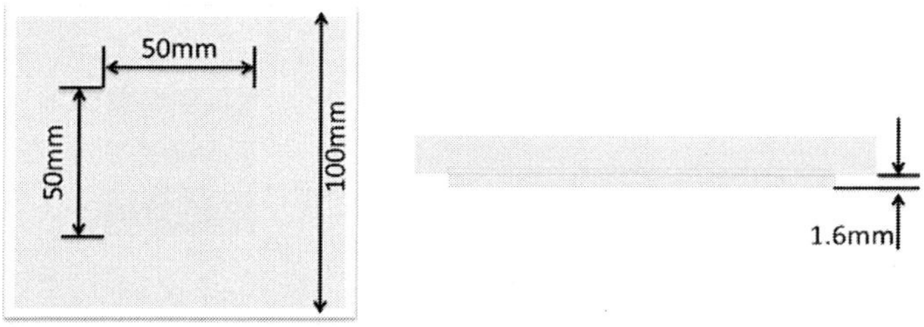

Figure 1. The geometry considered for numerical simulations.

When the computed solution was not changing significantly then the mesh with the minimal number of points has been selected. Due to the presence of strong gradients near the interface, and the necessity to refine the mesh close to it because of evaporation, the average number of computational points resulted to be 347,000 while the hexahedral control volumes accounted to 1,100,000. The general working framework of investigation consists of a layer of pure volatile liquid in contact with an inert gas flowing at different speeds. The gas is considered as ideal and incompressible. In each phase the physic-chemical properties and transport coefficient are considered constant. Those assumptions are valid if the evaporation rate and the thermal gradient are not too large. It is supposed that both the gas and the liquid layer are of fixed thickness; for the latter to be valid during the evaporation process, it is supposed that an equivalent amount of liquid is injected instantaneously to compensate for the evaporating liquid and that the fresh liquid injected does not alter the thermo-fluid dynamic field (vanishing Peclet number). Ethanol is considered as working fluid.

The set of equations solved in the gas and liquid bulk phase is the following:

$$\nabla \cdot \boldsymbol{V}_l = 0; \quad \nabla \cdot \boldsymbol{V}_g = 0 \tag{1.1}$$

$$\frac{\partial \boldsymbol{V}_l}{\partial t} + \boldsymbol{V}_l \cdot \nabla \boldsymbol{V}_l = \boldsymbol{V}_l \cdot \Delta \boldsymbol{V}_l - \frac{\nabla p}{\rho} \tag{1.2}$$

$$\frac{\partial \boldsymbol{V}_g}{\partial t} + \boldsymbol{V}_g \cdot \nabla \boldsymbol{V}_g = \boldsymbol{V}_g \cdot \Delta \boldsymbol{V}_g - \frac{\nabla p}{\rho} \tag{1.3}$$

$$\frac{\partial \boldsymbol{T}_l}{\partial t} + \boldsymbol{V}_l \cdot \nabla \boldsymbol{T}_l = \kappa_l \cdot \Delta \boldsymbol{T}_l \tag{1.4}$$

$$\frac{\partial \boldsymbol{T}_g}{\partial t} + \boldsymbol{V}_g \cdot \nabla \boldsymbol{T}_g = \kappa_g \cdot \Delta \boldsymbol{T}_g \tag{1.5}$$

$$\frac{\partial \boldsymbol{Y}_g}{\partial t} + \boldsymbol{V}_g \cdot \nabla \boldsymbol{Y}_g = D_g \cdot \Delta \boldsymbol{Y}_g \tag{1.6}$$

The system is closed by the following boundary conditions:

$$\boldsymbol{V} = \boldsymbol{0} \tag{1.7}$$

$$\mu_l\left(\frac{\partial V_x^l}{\partial y}\right)=\mu_l\left(\frac{\partial V_x^l}{\partial y}\right)-\sigma_T\cdot\frac{\partial}{\partial x}T_{\text{interface}}^l \tag{1.8}$$

$$\mu_l\left(\frac{\partial V_z^l}{\partial y}\right)=\mu_g\left(\frac{\partial V_z^l}{\partial y}\right)-\sigma_T\cdot\frac{\partial}{\partial z}T_{\text{interface}}^l \tag{1.9}$$

$$\hat{n}\cdot\boldsymbol{V}=0 \tag{1.10}$$

$$\hat{n}\cdot V=-\frac{D_g}{\left(1-Y_{\text{vap}}\right)}\cdot\frac{\partial Y_{\text{vap}}}{\partial\hat{n}} \tag{1.11}$$

$$\hat{n}\cdot\nabla T=0 \tag{1.12}$$

$$\lambda_g\hat{n}\cdot\nabla T_g=L_{\text{vap}}J_{\text{vap}}+\lambda_l\hat{n}\cdot\nabla T_l \tag{1.13}$$

$$T_g=T_l=T_{\text{interface}} \tag{1.14}$$

$$T_0=T_{\text{ref}}=293.15\text{ K} \tag{1.15}$$

$$\hat{n}\cdot\nabla Y=0 \tag{1.16}$$

$$Y_{\text{vap}}=r\cdot\frac{P_s\left(T_{\text{interface}}\right)}{P_{\text{total}}-P_s\left(T_{\text{interface}}\right)\cdot(1-r)} \tag{1.17}$$

$$Y_g=1-Y_{\text{vap}} \tag{1.18}$$

To deal with the interface between the two fluids and to appropriately implement the boundary conditions, the following guidelines were applied:

1) The interface between two different zones—no matter the kind of phases it separates (e.g. liquid/liquid or liquid/gas)—is considered as a two-sided wall without thickness whose normal are positives if directed outward.

2) Neither absorption of gas in the liquid phase is considered nor evaporation of the liquid. The gas channel is considered to be filled by pure inert gas, while the liquid cavity by pure liquid.

3) The continuity of the tangential velocity components of the interfacial velocity is also implemented, by equalizing at each calculation step the

velocity of the gas side of the interface to the one evaluated in the liquid from the shear stress conditions.

4) The Laplace pressure at the interface has been neglected by considering that the interface is flat and non- deformable. This assumption allows for greatly simplify the momentum conditions in direction normal to the interface.

3. RESULTS

The evolution of the system has been studied respect to variations of the volume flow rate of the gas shear flow. Three flow rates have been investigated: 1 L/min, 4 L/min and 8 L/min. The corresponding average velocities as computed at the outlet section range between 0.04 - 0.33 m/sec, while Reynolds numbers vary between 14 and 115, the laminar character of the flow being confirmed.

One of the goals of the CFD analysis was to follow the evolution of the thermal patterns generated at the interface as result of the interaction between thermo-capillary stresses and shear flow ones. Figure 2 shows the evolution of the thermal patterns for a volumetric flow rate of 1 L/min. Pictures are taken at interval of approximately 15 seconds. It is possible to notice that at the beginning of the computation thermal patterns size is approximately 1/10 of the interface length (Figure 2(a)) and it is slowly increasing during the evolution of the interfacial thermal field. After 50 sec., the typical size of the thermal cells ranges between 1/5 and 1/10 of the interface length. The computed maximal temperature difference at the interface, i.e. the temperature difference between the coldest and hottest point at the interface is ~2.8 K.

An analogous trend could be spotted in the case of an imposed volumetric flow rate of 4 L/min. In this case, after the initial transient leading to cells formation and coalescence a quasi-steady state is reached with a typical thermal cells size of 1/4 of the interfacial length (Figure 3). The computed interfacial temperature difference accounts for ~3.9 K.

In the latter case studied, volumetric flow rate of 8 L/min (Figure 4), the transient phase is characterized by the formation of two rows of elongated cells on the lateral border of the interface. They coalesce in larger thermal cells having a characteristic size ranging between 1/3 and 1/5 of the interfacial length. In this latter case, the computed temperature difference at the interface is ~4.2 K.

In all cases, some common features appear.

- The size of the thermal cells is increasing due to the thermal gradient relaxation in the bulk liquid. In the transient phase, the liquid layer acts as a layer cooled from above. The thermal boundary layer in the liquid bulk phase increases by thermal diffusion during the evolution of the system and as a consequence, being the characteristic number of the thermal cells approximatively proportional to the thermal gradient thickness in the bulk phase, the typical size of the thermal cells increases [14] .
- The distribution of temperature shows a minimum at the inlet border. Along the inlet line—the line where the gas flows meet the evaporating liquid—the gas has the maximal efficiency in removing vapor from the interface. The inlet line corresponds to the line where the evaporation is maximal and the cooling effects more efficient.
- In Figure 5 and Figure 6, the evaporation rate time evolution is plotted for the different gas flow rates studied. It is interesting to point out that:
- The evaporation rate variation has a rapid, strong transient in a time lag of few seconds (Figure 6). At the beginning no vapor is present in the gas phase. The higher values of the evaporation rate could be explained by considering that the entropic driving force for the evaporation process to appear is strongest when the liquid is evaporating in a vapor-free gas layer.
- The evaporation rate is increasing with increasing volumetric gas flow rate. However, this increase is not linearly proportional. More in details, the evaporation rate increase for a fourfold increase in the gas flow rate accounts only to 2.5 times. The increase is further reduced when passing to eightfold gas flow rate.

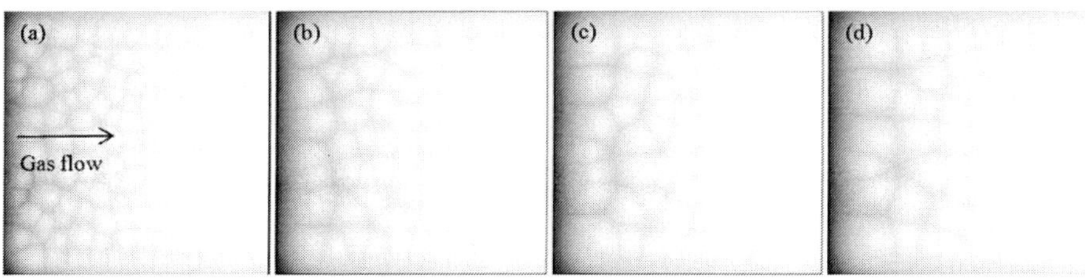

Figure 2. Thermal patterns at the interface for a volumetric gas flow of 1 L/min: (a) 1 sec; (b) 15 sec; (c) 30 sec; (d) 50 sec.

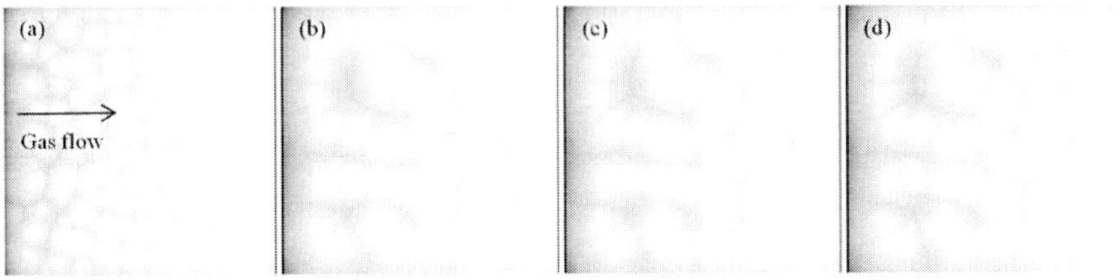

Figure 3. Thermal patterns at the interface for a volumetric gas flow of 4 L/min: (a) 1 sec; (b) 15 sec; (c) 30 sec; (d) 50 sec.

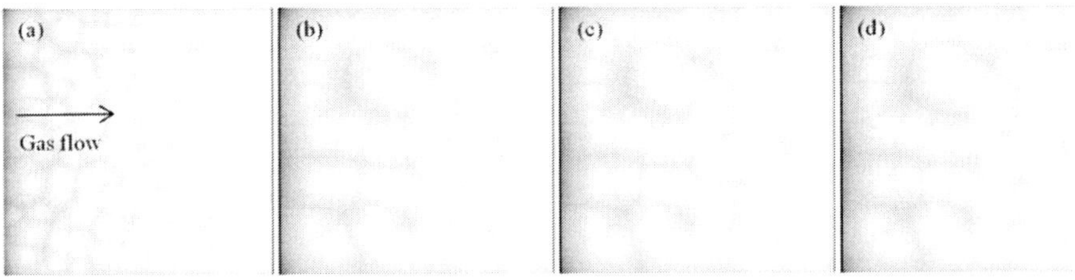

Figure 4. Thermal patterns at the interface for a volumetric gas flow of 8 L/min: (a) 20 sec; (b) 40 sec; (c) 60 sec; (d) 80 sec.

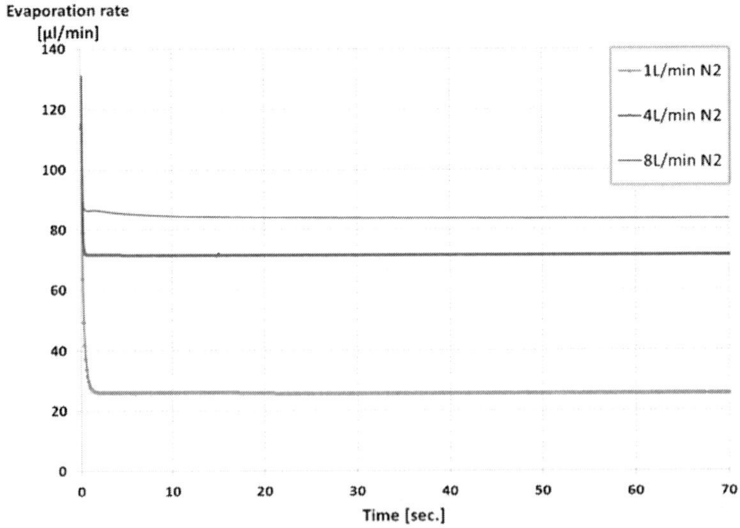

Figure 5. Evaporation rate as a function of N2 flow rate.

- After the transient, the evaporation rate stabilizes at a quasi-steady value. This point is rather meaningful because the thermal patterns at the interface are continuing to evolve while keeping the total evaporation rate constant.

To better understand this latter point, concentration boundary layers at the center of the interfacial region have been computed and compared. Figures 7-9 show that, for all cases, the concentration boundary layer in the gas phase evolves toward a constant and steady distribution after a short transient, approximatively 2 sec. for a volumetric gas flow rate of 1 L/min, less than 1 sec. for a gas flow of 4 L/min and less than 0.5 sec. for 8 L/min.

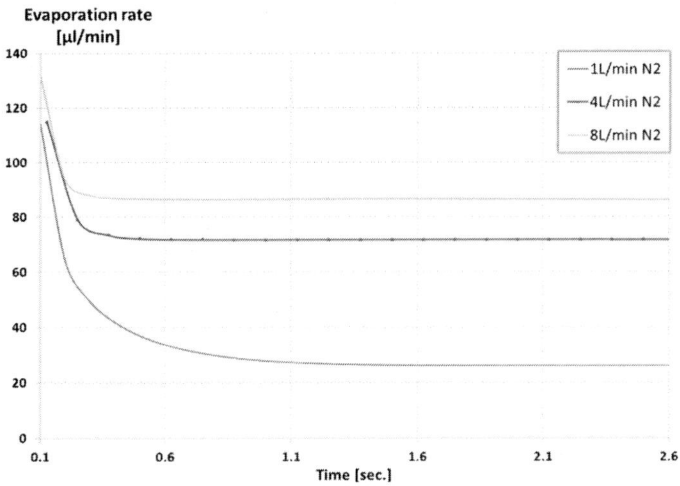

Figure 6. Evaporation rate as a function of N2 flow rate—initial transient.

By comparison with Figure 6, it could be inferred that the main driving force in the evolution of the evaporation rate at the interface is the creation and stabilization of the vapor boundary layer [15] . Evaporation from the interface is, in these conditions, diffusion limited by the efficiency of the gas flow in removing vapor from the interface.

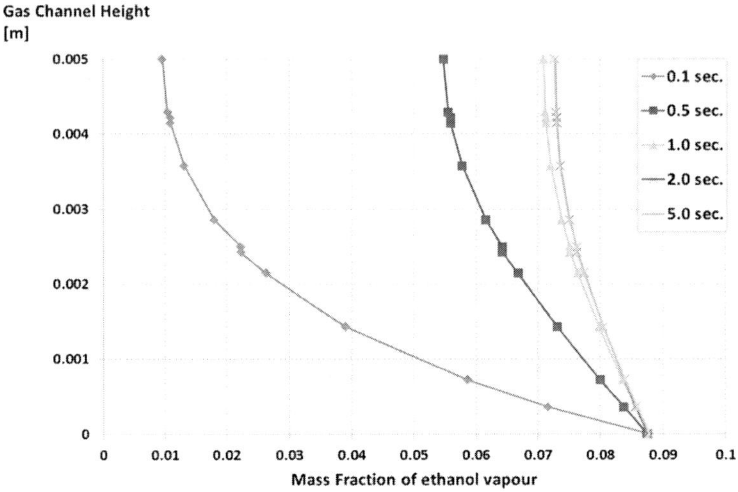

Figure 7. Concentration boundary layer evolution for a volumetric flow rate of 1 L/min.

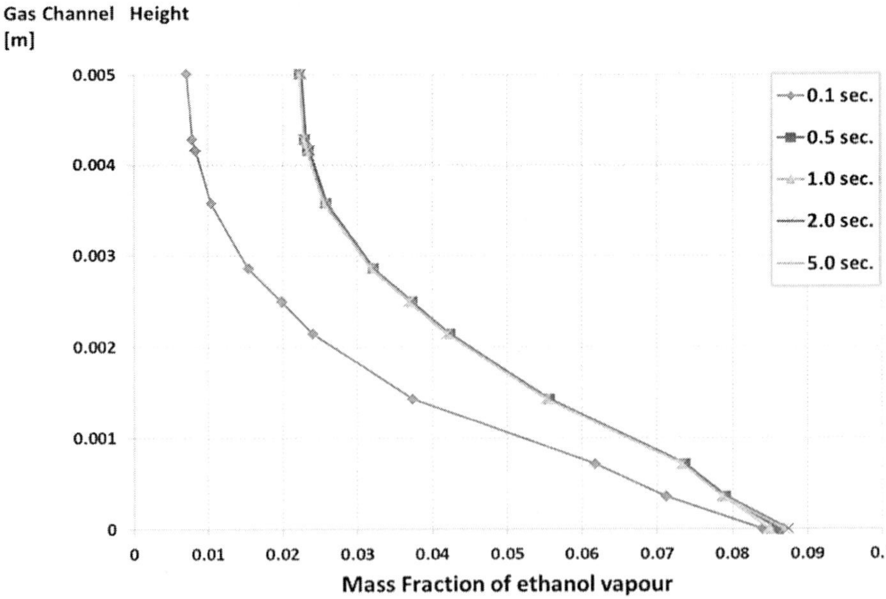

Figure 8. Concentration boundary layer evolution for a volumetric flow rate of 4 L/min.

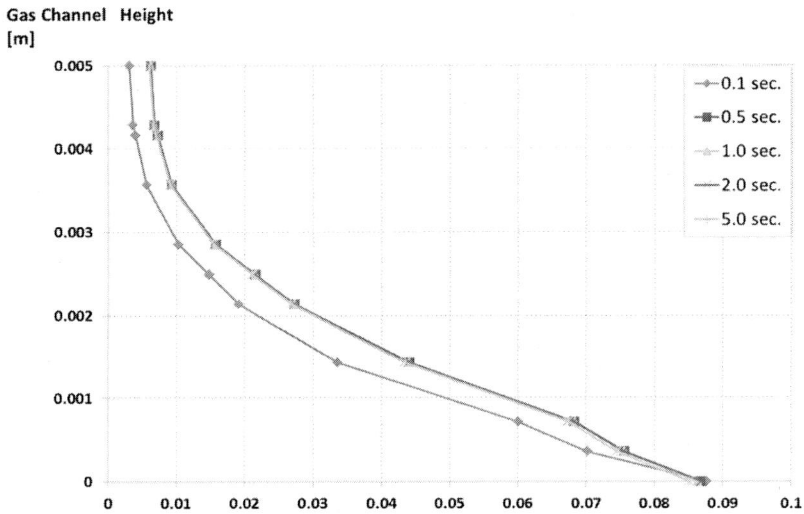

Figure 9. Concentration boundary layer evolution for a volumetric flow rate of 8 L/min.

4. CONCLUSION

The coupling between evaporation and thermo-capillary convection has a strong effect on the thermo-fluid-dy- namic field of a liquid layer subject to a flow of inert gas. Common features have been spotted concerning the evolution of the thermal patterns: the size dependence on the relaxation of the thermal gradient in the bulk liquid phase and the typical size of the thermal spot. Also, the evaporation rate is strongly affected by the vapour distribution in the gas phase. When the vapour boundary layer has reached a given steady state, the total evaporation rate is only barely changing by the evolution of the thermal field at the level of the interface. This result could be of importance in the designing and optimization of enhanced evaporators for cooling of electronic and power devices.

ACKNOWLEDGEMENTS

The authors would like to thank the ESA Microgravity Application Program dedicated to the evaporation phenomena and applications for the financial support in achieving the presented results.

REFERENCES

1. Nie, Z.H. and Kumacheva, E. (2008) Patterning Surfaces with Functional Polymers. Nature Materials, 7, 277-290.
2. Scheid,B., Margerit, J., Iorio, C.S., Joannes, L, Heraud, M., Queeckers, P.,Dauby, P. C. and Colinet, P. (2012) Onset of Thermal Ripples at the Interface of an Evaporating Liquid under a Flow of Inert Gas. Experiments in fluids, 52, 1107- 1119.
3. Bestehorn, M. and Merkt, D. (2006) Regular Surface Patterns on Rayleigh-Taylor Unstable Evaporating Films Heated from Below. Physical Review Letters, 97, Article ID: 127802.
4. Molenkamp, T. (1998) Marangoni Convection, Mass Transfer and Microgravity. Ph.D. Dissertation, Rijksuniversiteit Groningen, Groningen.
5. Liu, R. and Kabov, O.A. (2012) Instabilities in a Horizontal Liquid Layer in Cocurrent Gas Flow with an Evaporating Interface. Physical Review E, 85, Article ID: 066305.
6. Andreev, V.K. and Bekezhanova, V.B. (2013) Stability of Non-Isothermal Fluids (Review). Journal of Applied Mechanics and Technical Physics, 54, 171-184.
7. Mancini, H. and Maza, D. (2004) Pattern Formation without Heating in an Evaporative Convection Experiment. Europhysics Letters, 66, 812-818.
8. Haut, B. and Colinet, P. (2004) Surface-Tension-Driven Instabilities of a Pure Liquid Layer Evaporating into an Inert Gas. Journal of Colloid and Interface Science, 285, 296-305.
9. Goncharova, O. and Kabov, O. (2010) Mathematical and Numerical Modeling of Convection in a Horizontal Layer under Co-Current Gas Flow. International Journal of Heat and Mass Transfer, 53, 2795-2807.
10. Goncharova, O.N. and Kabov, O.A. (2009) Numerical Modeling of the Tangential Stress Effects on Convective Fluid Flows in an Open Cavity. Microgravity Science and Technology, 21, S119-S127.
11. Iorio, C.S., Goncharova, O.N. and Kabov, O.A. (2011) Heat and Mass Transfer Control by Evaporative Thermal Patterning of Thin Liquid Layers. Computational Thermal Sciences, 3, 333-342.

12. Grace, S.M., Dewar, W.G. and Wroblewski, D.E. (2004) Experimental Investigation of the Flow Characteristics within a Shallow Wall Cavity for both Laminar and Turbulent Upstream Boundary Layers. Experiments in Fluids, 36, 791- 804.

13. (2006) Fluent User Guide Manual. Fluent Inc., Cecil Township.

14. Colinet, P., Legros, J.C. and Velarde, M.G. (2001) Nonlinear Dynamics of Surface-Tension-Driven Instabilities. Wiley-VCH, Berlin.

15. Kanatani, K. (2013) Effects of Convection and Diffusion of the Vapour in Evaporating Liquid Films. Journal of Fluid Mechanics, 732, 128-149.

2

CHAPTER

NUMERICAL STUDY OF THERMAL PERFORMANCE OF A CAPILLARY EVAPORATOR IN A LOOP HEAT PIPE WITH LIQUID-SATURATED WICK

Masahito Nishikawara[1], Hosei Nagano[1], Laetitia Mottet[2], Marc Prat[2]

[1]*Department of Aerospace Engineering, Nagoya University, Nagoya, Japan*

[2]*INPT, UPS, IMFT (Institute of Fluid Mechanics of Toulouse), Toulouse University, Toulouse, France*

ABSTRACT

Heat transfer of a capillary evaporator in a loop heat pipe was analyzed through 3D numerical simulations to study the effects of the thermal conductivity of the wick, the contact area between the casing and the wick, and the subcooling in the compensation chamber (CC) on the thermal performance of the evaporator. A pore network model with a distribution of pore radii was used to simulate liquid flow in the porous structure of the wick. To obtain high accuracy, fine meshes were used at the boundaries among the casing, the wick, and the grooves. Distributions of temperature, pressure, and mass flow rate were compared for polytetra-fluoroethylene (PTFE) and stainless steel wicks. The thermal conductivity of the wick and the contact area between the casing and the wick significantly impacted thermal performance of the evaporator heat-transfer coefficient and the heat leak to the CC. The 3D analysis provided highly accurate values for the heat leak; in some cases, the heat leaks of PTFE and stainless steel wicks showed little differences. In general, the heat flux is concentrated at the boundaries between the casing, the wick, and the grooves; therefore, thermal performance can be optimized by increasing the length of the boundary.

Keywords: Capillary Evaporator, Loop Heat Pipe, Numerical Simulation, Pore Network Model, Two-Phase Heat Transfer

1. INTRODUCTION

Loop heat pipes (LHPs) and capillary pumped loops (CPLs) have attracted attention as advanced thermal control devices of spacecrafts, electronic devices, etc. The LHP and CPL require no external power to drive working fluids because of the capillary pressure generated in a porous structure in the evaporator, and they have high heat-transfer capabilities because heat transport is associated with phase changes between liquid and vapor (Figure 1(a)). Operating principles and characteristics are explained in detail elsewhere [1] [2] .

The thermal performance of these devises is significantly affected by the design of the evaporator, and several studies of heat and mass transfer in capillary evaporators have been reported [3] -[13] . Fluid in the wick takes one of two principal forms: at low heat loads, the wick is completely filled with liquid, whereas at high heat loads, the wick contains fluid in a two-phase liquid-vapor state. Since it is complicated to propose an optimum geometry of the evaporator for the two widely different conditions, this paper focuses on the first condition. Numerical studies on heat transfer with saturated wicks have been performed [3] -[9] . In [5] , a quasi-3D numerical model was developed on the basis of Darcy's law and energy conservation in the evaporator, including liquid in the compensation chamber (CC), the wick, the groove, and the casing; the model was used to obtain distributions of temperature, pressure, and velocity. In [6] , a 3D simulation using the lattice Boltzmann method was presented, and details of flow and heat transfer in the porous structure were obtained.

A high-performance evaporator requires high-efficiency heat exchange between the casing of the heating surface and the vapor transferred to the condenser. In addition, heat leaks to the CC must be small because the saturated state in the CC controls the state of the LHP. Therefore, evaporator designs for thermal performance can be discussed in terms of the trade-off between the heat-transfer coefficient and heat leaks to the CC. This approach has been addressed in [14] . In our previous study, polytetrafluoroethylene (PTFE) wicks, with a bulk thermal conductivity of 0.25 W/m·K, were applied in an attempt to decrease heat leaks to the CC [8] [15] [16] . Moreover, there is a requirement also for the evaporator to transfer the maximum heat load. To increase the heat-transfer amount reaching the capillary limit, the reduction of the pressure loss by the fluid flow should be considered. However, this point is not included in this paper.

In this study, a 3D numerical model of an LHP evaporator with the wick fully saturated with liquid is developed. The purpose was to find a guide for a design that attains geometric optimization in terms of the capillary evaporator's thermal performance. Focus has been provided on how the thermal performance of the evaporator is affected by the wick's thermal conductivity, the contact area between the casing and the wick, and the subcooling in the CC. Note that, in this study, situations wherein the wick contains fluid in two-phase vapor-liquid states are not considered; those situations have been addressed elsewhere [10] -[12] . In Section 2, the governing equations and the porous characteristics of the wick are described. In Section 3, our numerical procedure is discussed. Finally, in Section 4, computational results are presented. Comparison of PTFE with stainless steel as a representative material of metal wick is discussed specially.

2. NUMERICAL MODEL

This section presents our numerical model for heat and mass transfer in the evaporator. Figure 1(b) shows the computational domain, which corresponds to the area surrounded by the red dashed line in Figure 1(a). Thewick contains grooves, as observed in Figure 1(b). The pore network model (PNM) in [9] [11] [12] was applied here to determine how liquid flow in the wick is affected by the pore radius distribution in the porous media.

Figure 1(c) shows the PNM. The PNM divides the porous structure into two parts-pores and throats. In our case, the pores were spherical and the throats were cylindrical. The radii of the pores and throats together with the lengths of the throats characterize the porous structure of the wick, which includes pore radius, porosity, and

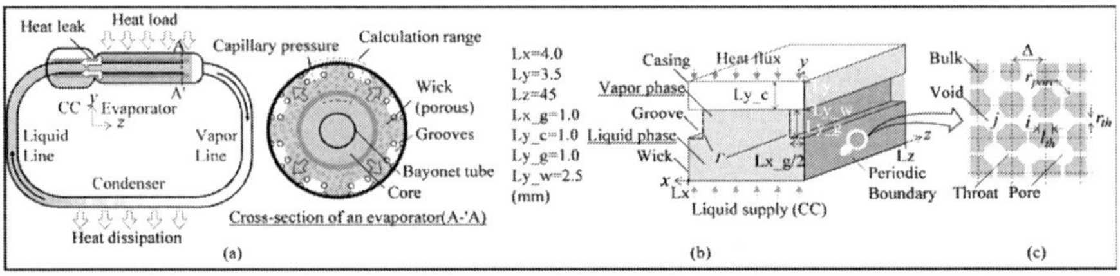

Figure 1. (a) Schematic of a loop heat pipe, (b) computational domain, and (c) representation of pore network model for fluid analysis in the wick.

permeability. Using this simple model, the PNM can include more pores and calculate over a larger porous media than the lattice Boltzmann method. Energy conservation in other components of the LHP was excluded, allowing us to easily determine heat-transfer only in the evaporator.

The main assumptions used in the model were as follows: 1) grooves were filled with saturated vapor at constant temperature and pressure; 2) the bulk solid and fluid in the wick were in local thermal equilibrium; 3) the fluid was incompressible; 4) fluid properties were temperature dependent; 5) the liquid-vapor interface had no thickness; 6) the process was at steady state; and 7) the effects of gravity and thermal radiation were negligible.

2.1. Governing Equation

In the PNM, mass flow rate $\dot{m}_{ij}$ in the throat between pores i and j is expressed as follows,

$$\dot{m}_{ij} = \frac{g_{ij}}{v_l}\left(P_i - P_j\right) \tag{1}$$

where, g_{ij} is flow resistance in the throat, v_l is kinematic viscosity coefficient of liquid, P is pressure in a pore. Assuming Poiseuille flow, the flow resistance is given by:

$$g_{ij} = \frac{\pi r_{th}^4}{8 l_{th}} \tag{2}$$

where, r_{th} and l_{th} are radius and length of the throat, respectively. By considering mass conservation at pore i, system of linear equations is obtained,

$$\sum_j \frac{g_{ij}}{v_l}\left(P_i - P_j\right) = 0 \tag{3}$$

Energy conservation in the wick, including conduction and convection, can be written as

$$c_{p\_l}\nabla\left(\dot{m}_{ij}T\right) = k_{\text{eff}\_l}\nabla^2 T \tag{4}$$

and heat conduction by the evaporator casing is given by

$$k_{\text{case}}\nabla^2 T = 0 \tag{5}$$

In Equations (4) and (5), c_p is the isobaric specific heat of the liquid, T is temperature, k_{case} is thermal conductivity of the casing, and k_{eff_l} is the effective thermal conductivity of the wick filled with liquid. The latter was obtained by volume averaging the thermal conductivities of the bulk solid k_{bulk} and liquid $k_{\_l}$ each appropriately weighted with the porosity ε :

$$k_{\text{eff}\_l} = \varepsilon k_{\_l} + \left(1-\varepsilon\right)k_{\text{bulk}} \tag{6}$$

Liquid evaporate at the interface between the wick and the grooves, which is defined as Γ interface. The Γ interface is shown in Figure 2. The mass flow rate generated by evaporation $\dot{m}_n$ is expressed by

$$-k_{\text{eff}\_l}\left.\frac{\partial T}{\partial n}\right|_{\Gamma} = h_i\left(T - T_{gr}\right) \tag{7}$$

$$h_i A\left(T - T_{gr}\right) = \dot{m}_n H_{fg} \tag{8}$$

where n is the direction normal to the Γ interface, h_i is the interfacial heat-transfer coefficient calculated by using the equation in [17] , T_{gr} is the temperature of the grooves, A is the cross-sectional area, and H_{fg} is the latent heat of the working fluid. The temperature in the grooves was assumed to be constant at the saturation temperature. The saturation temperature was calculated from the pressure in the grooves and the P-T saturation curve of the working fluid. The pressure loss from the grooves to the CC ΔP_{lines} was calculated from the total mass flow rate in the wick, assuming Poiseuille flow in a circular pipe, as presented in [18] . The pressure in the grooves was obtained from

$$P_{gr} = P_{cc} + \Delta P_{\text{lines}} \tag{9}$$

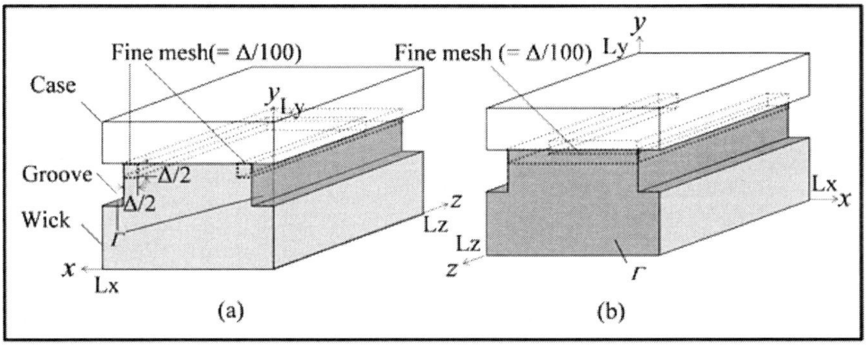

Figure 2. Liquid-vapor interface (Γ interface, orange area) and region of fine mesh (volime in dotted line). (a) front view and (b) rear view.

where P_{cc} is the pressure in the CC. The temperature and pressure in the CC were not calculated in this study. The assumptions of constant temperature and saturation pressure were imposed to allow simple calculations of only the heat transfer in the evaporator. The other boundary conditions were those commonly used in previous studies: 1) uniform heat flux at the heating surface $(y = Ly)$ (Neumann boundary condition); 2) thermal connection by contact thermal resistance of 10,000 W/m²·K used in [19] at the casing-wick contact surface $(y = Ly\_w)$; 3) convective boundary at the bottom of the wick $(y = 0)$ with a heat-transfer coefficient of 100 W/m²·K used in a previous study [13] (Robin boundary condition); 4) constant pressure P_{cc} at the bottom of the wick $(y=0)$ (Dirichlet condition); 5) impervious walls at $z=0$ and $y=Ly\_w$; 6) adiabatic conditions at $z=0$ and at the casing-groove interface; and 7) periodic boundary conditions at $x=0$ and $x=Lx$.

2.2. Representation of Porous Structure

In the PNM, characteristics of the porous structure can be established by setting dimensions for voids—the radius of each pore r_{pore}, the radius of each throat r_{th}, and the length of each throat l_{th} (see Figure 1(c)) —which control the porosity, permeability, and pore radius distribution, respectively. In this study, a polytetrafluoroethylene (PTFE) wick [16] , for which the porosity and permeability are 0.34 and 2.0 × 10^{-14} m^2, respectively, is modeled. The mode and maximum pore radius were 1.2 and 2.1 μm, respectively. The throat radius was determined by using a log-normal probability density function to fit the pore radius distribution, which was measured by mercury porosimetry [20] . A mercury porosimeter can measure a pore radius distribution when the capillary pressure balances. In the PNM, the capillary pressure is expressed in terms of the throat's radius. Therefore, in the PNM, the measured pore radius distribution corresponds to the radius distribution of the throat. Figure 3 compares measured and calculated distributions of pore radii. In the figure, the y-axis contains the pore volume normalized by its maximum. The distribution of throat radii agrees well with the measured distribution of pore radii. To determine the permeability of the constructed wick, the permeability was calculated as introduced in [12] . The computed permeability agreed with the measured permeability within 2%. All the main porous characteristics were in good agreement with measured values. The representation of porous characteristics is presented in [9] . Once the porous structure was constructed, it was used in all calculations.

3. NUMERICAL PROCEDURE

The method used to solve the above equations is shown by the flow chart in Figure 4. First, standard finite volume methods were used to solve the combination of Equations (4) and (5) with no mass flow rate. From those results, the mass flow rate at the Γ interface was calculated, as imposed by the boundary condition in Equation (3). After the pressure field was obtained from Equation (4), mass flow rates in the wick and the LHP, and the pressure drop from the grooves to the CC were calculated. Finally, the updated pressure drop and mass flow rate in the wick were compared with values from previous iterations to judge convergence of the calculation. If convergence was not yet attained, the calculation returned to Equations (4) and (5) with the updated mass flow rate. The above calculation loop was iterated until convergence was obtained.

An energy balance over the entire system can be written as

$$\dot{Q}_{\text{leak}} + \dot{Q}_{\text{evap}} + \dot{Q}_{\text{sens}} = \int_{y=Ly} \dot{q}_{\text{apply}} \mathrm{d}A \tag{10}$$

where, $\dot{Q}_{\text{leak}}$ is heat leak to the CC, $\dot{Q}_{\text{evap}}$ is heat transferred by phase change, and $\dot{Q}_{\text{sens}}$ is sensible heat from the inlet to the outlet of the wick. The right side is total amount of applied heat to the evaporator. Each term in the left side is calculated as follows:

$$\dot{Q}_{\text{leak}} = \int_{y=0} \left(-k_{\text{eff}\_l} \frac{\partial T}{\partial y} \right) \mathrm{d}A \tag{11}$$

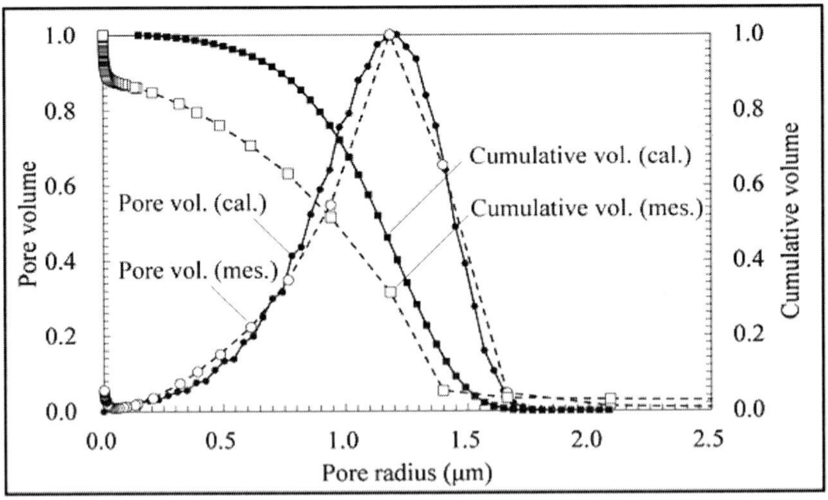

Figure 3. Comparison of the measured and calculated pore radius distributions. Vertical axes are normalized by the maximum value.

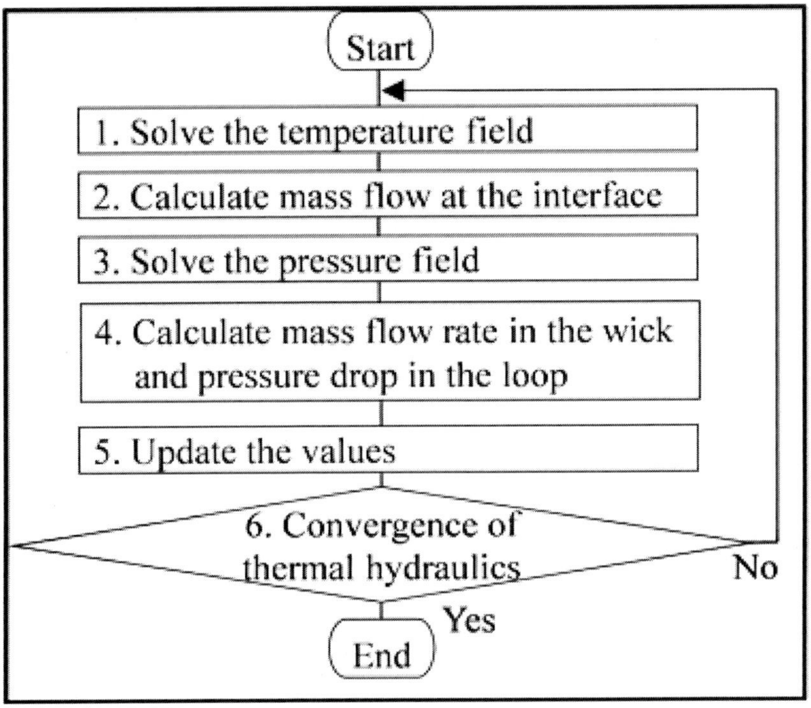

Figure 4. Flow chart of the numerical algorithm.

$$\dot{Q}_{evap} = \sum_{\Gamma} \dot{m}_n H_{fg} \tag{12}$$

$$\dot{Q}_{sens} = \sum_{\Gamma} \dot{m} c_{p\_l} T - \sum_{y=0} \dot{m} c_{p\_l} T \tag{13}$$

The energy balance (Equation (10)) was satisfied with a maximum relative error of less than 1%.

Generally, a mesh size of $\Delta = 0.1$ mm is used. However, in case the thermal conductivity of the wick is lower than that of the casing, the temperature gradient in the wick near the boundary among the casing, the wick, and the grooves can be high. Therefore, to calculate with a high degree of accuracy, a finer mesh of $\Delta/100$ is used, as shown in Figure 2. The total number of nodes was 10,773,391.

Table 1. Calculation conditions using either PTFE or stainless steel wicks.

Calculation number	1	2	3	4	5	6	7	8
Material of a wick	PTFE[a]	PTFE	PTFE	SUS[b]	SUS	SUS	PTFE	SUS
Area of contact surface (%)	75	50	25	75	50	25	50	50
Subcooling of the CC (°C)	0	0	0	0	0	0	5	5

[a]Thermal conductivity is 0.25 W/m·K. [b]SUS is stainless steel and the thermal conductivity is 16 W/m·K.

4. RESULTS AND DISCUSSION

Calculations were performed for the eight conditions shown in Table 1. The conditions are selected to investigate how the thermal performance of the evaporator was affected by the thermal conductivity of wicks, contact area between wick and casing, and the subcooling of liquid in the CC. The effective thermal conductivities of the PTFE and stainless steel wicks completely filled with liquid were 0.22 and 10.6 W/m·K, respectively. The contact area between the casing and the wick depends on the width of the grooves Lx_g; in Table 1, the contact area is expressed as the percentage of casing cross-sectional area, $Lx \times Lz$. Subcooling in the CC was used as the liquid temperature in the CC, which was the boundary condition at $y = 0$. Other conditions included the following: the heat flux applied to the casing was 0.625 W/cm^2, the saturation temperature in the CC was 29.4°C, and the thermal conductivity of the stainless steel casing was 16 W/m·K. The working fluid was ethanol.

Figure 5 shows 3D color renderings for pressure and temperature distributions and 2D color renderings of mass flow rate distributions on the x-y plane at $z = Lz/2$ for the PTFE (left) and stainless steel (right) wicks; these results are from calculations 2 and 5 in Table 1. Figure 5(a) shows little difference in the pressure distributions: the maximum pressure was in the grooves, and the minimum pressure was right under the evaporator case at the groove-wick interface. The maximum pressure difference was about 3.5 kPa, which was much less than the capillary pressure (18 kPa) for the largest pore radius. This confirms that it was reasonable to assume that the liquid-vapor interface was at the interface between the wick and the grooves.

In Figure 5(b), the maximum temperature was on top of the casing, and the minimum temperature was at the lowest part of the wick. The casing temperature with the PTFE wick was higher than that with the stainless steel wick. In the model, the saturation temperature in the CC was set directly to study the performance of only the evaporator. In fact, the CC temperature is affected by the heat leak from the evaporator and the subcooled liquid flowing from the liquid line. Therefore, the CC temperature affects the casing temperature, and evaluation of the temperature in the calculation differs somewhat from reality. Note that the color scales are different for the two panels in Figure 5(b). Note also that, although the boundary conditions on the front and rear surfaces differ, there is almost no variation of temperature or pressure in the z-direction in Figures 5(a)-(b).

In Figure 5(c), the two distributions of mass flow rate are very different. For the PTFE wick, the maximum mass flow rate is at the boundary between the casing, the wick, and the grooves, and the variations are large. In contrast, for the stainless steel wick, the variations are small. It seems that the mass flow rate in the PTFE wick is more sensitive to the temperature distribution.

Figure 6 presents calculated results for the response of the evaporator heat-transfer coefficient to changes in contact areas when PTFE and stainless steel wicks were used; these results are from calculations 1 - 6 in Table 1.

The evaporator heat-transfer coefficient h_{evap} was calculated from

$$h_{\mathrm{evap}} = \frac{\dot{Q}_{\mathrm{evap}}}{A_{\mathrm{cont}}\left(T_{\mathrm{case}} - T_{gr}\right)} \tag{14}$$

where A_{cont} is the contact area between the case and wick, and T_{case} is the average temperature on the upper surface of the casing $(y = Ly)$. Here, the performance of the evaporator, without the effect of the heat leak to the CC, was evaluated using $\dot{Q}_{\mathrm{evap}}$ as the amount of heat instead of using the heat load to the evaporator. Figure 6 shows that the evaporator heat-transfer coefficient with the stainless steel wick was approximately 10 times higher than that with the PTFE wick. This indicates that the thermal conductivity of the wick has a large impact on the heat-transfer coefficient of the evaporator. The figure also shows that, for both wicks, the evaporator heat-transfer coefficient increases as the contact area decreases.

Figure 7 presents distributions of heat flux in the y-direction on the contact surface at $z = Lz/2$ for the PTFE and stainless steel wicks; these are results from calculations 2 and 5 in Table 1. Heat flux in negative y-direction is positive along the vertical axis. For both wicks, the maximum heat fluxes occurred at the groove-wick interface, which were at $x = 1.0$ and 3.0 mm. The maximum heat flux for the PTFE wick was 12 times larger than that for the stainless steel wick. At the groove-wick interface ($x = 1.0$ and 3.0 mm), high heat flow occurs in the x-direction due to evaporation with a high interfacial heat-transfer coefficient. As a result, the between the casing and the wick at $z = Lz/2$ for PTFE and stainless steel wicks.

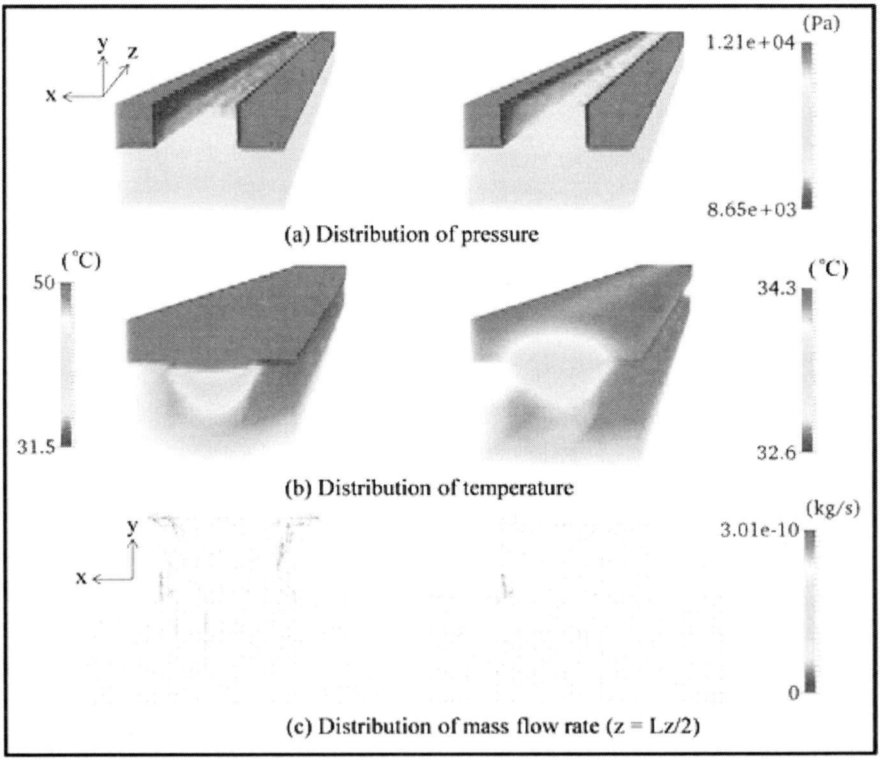

Figure 5. Renderings of (a) 3D pressure, (b) 3D temperature, and (c) 2D mass flow rate distributions for PTFE wick (left, calculation 2) and for stainless steel wick (right, calculation 5).

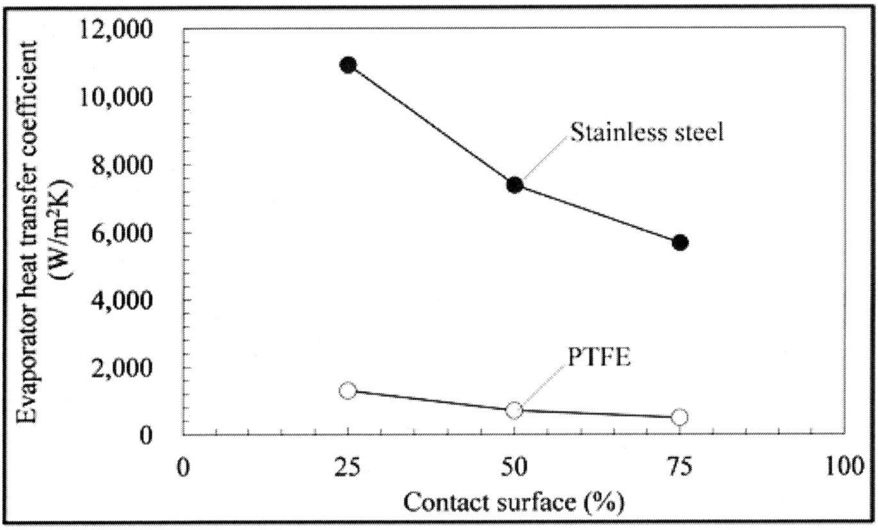

Figure 6. Effect of contact area on evaporator heat-transfer coefficients for PTFE and stainless steel wicks.

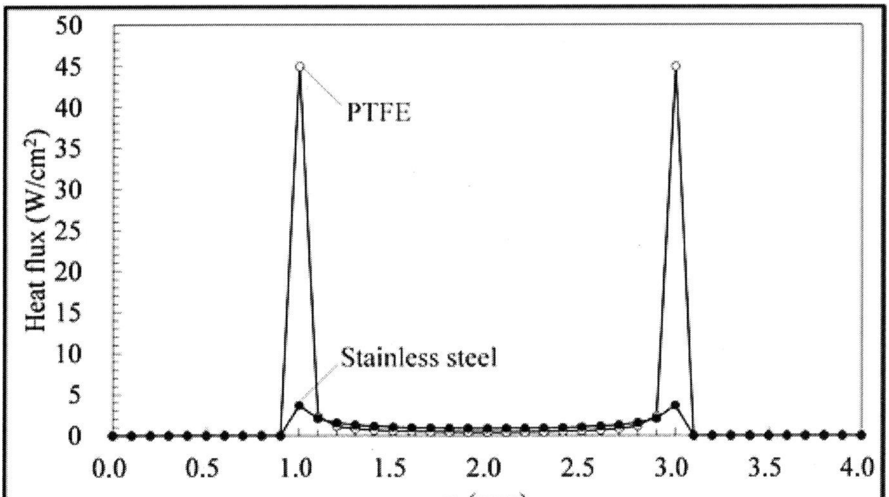

Figure 7. Comparison of heat flux distribution along the contact surface

heat flux concentrates at the boundaries between the casing, the wick, and the grooves. Therefore, the evaporator heat-transfer coefficient is higher when the region of the boundary is larger. Therefore, the thermal performance of the evaporator can be optimized by changing the length of the boundary per unit area of contact surface. In this calculation, the lengths were 0.67, 1.0, and 2.0 mm on contact areas of 75%, 50%, and 25%, respectively. In [8] [21] , increasing the number of grooves, i.e., increasing of the length of the boundary, led to enhance the thermal performance of the evaporator. The pressure loss in the grooves must also be included when considering the LHP performance. The pressure loss depends on groove length and cross-section area, since it can be estimated from flow in a pipe. Thus, geometric optimization of the evaporator for LHP performance can be discussed in terms of the trade-off between the length of the boundary and the geometrical parameters affecting the pressure loss. The optimization theory based on the above will be developed and be verified by LHP experiments in a future work.

Table 2 lists the calculated results at each calculation condition for the evaporator heat-transfer coefficient h_{evap}, the heat transferred by phase change $\dot{Q}_{evap}$, and the heat leak to the CC $\dot{Q}_{leak}$. The two amounts of heat were normalized using the heat load to the evaporator from the rhs of Equation (10). Although the effective thermal conductivities of the two wicks differed by a factor of 48, the heat leaks for the PTFE and stainless steel wicks were almost the same (see results for calculations 1 and 4 in Table 2). Based on an estimate of the heat leak assuming one-dimensional heat conduction in the wick [22] -[25] , the heat leak of the PTFE wick, which has a lower thermal conductivity, is smaller. However, results from the two calculations do not correspond. On the other hand, on data comparison between calculation numbers 3 and 6, the results still provide valid estimations. In [25] , the convection effect of liquid flow in the wick is small because the Peclet number was 0.4. The heat leak can generally be estimated with higher accuracy by 3D analysis; thus, estimates of the heat leak assuming one-dimensional heat conduction in the wick can be less accurate in some cases. The reason may be the temperature distribution in the x-direction, as shown in Figure 5(b), and the formation of a region of superheated liquid in the wick. To estimate the heat leak with high accuracy, it will be necessary to develop a model that considers the above or to solve for the temperature distribution in the wick. In addition, when the entire LHP system evaluated, the heat leak to the CC through the metallic casing must to be considered.

Evaporator heat-transfer coefficients calculated with 5° of subcooling (calculations 7 and 8 in Table 2) were smaller than those without subcooling (calculations 2 and 5). This is because the heat leak $\dot{Q}_{\text{leak}}$ was larger due to the lower temperature of the CC by the subcooling; thus, the heat transferred by phase change $\dot{Q}_{\text{evap}}$ was smaller.

Figure 8 presents the distribution of mass flow rate in the wick?groove interface, as obtained from calculation 7. For this calculation, the length Lz was changed to 1.5 mm , which allows us to more easily view the entire evaporator. On surfaces having heights less than the grooves, the arrows representing mass flow rates in Figure 8 cannot be seen because they point into the wick. This implies that condensation occurred at the interface; this is a heat-pipe effect and has been recognized earlier in [26] . The total mass flow rate by condensation was 8.48% of the total mass flow rate in the evaporator. The effect on the PTFE wick was more noticeable than that on the stainless steel wick. It seems that the heat-pipe effect was larger for the PTFE wick because of the larger temperature distribution in the PTFE wick.

5. CONCLUSION

The thermal performance of the evaporator in a loop heat pipe using 3D numerical simulations with a pore network model for situations wherein the wick was fully filled with liquid, has been investigated. In particular, how the evaporator's performance responded to changes in the thermal conductivity of the wick, the contact area

Table 2. Calculated results of evaporator performance at each of the eight conditions listed in Table 1.

Calculation number	1	2	3	4	5	6	7	8
h_{evap} (W/m²·K)	497	714	1300	5680	7400	10900	678	6810
Q_{evap} (%)	91.2	92.9	93.8	93.5	93.8	93.8	87.8	85.0
Q_{leak} (%)	5.64	3.91	2.86	5.59	5.27	5.05	7.99	12.7

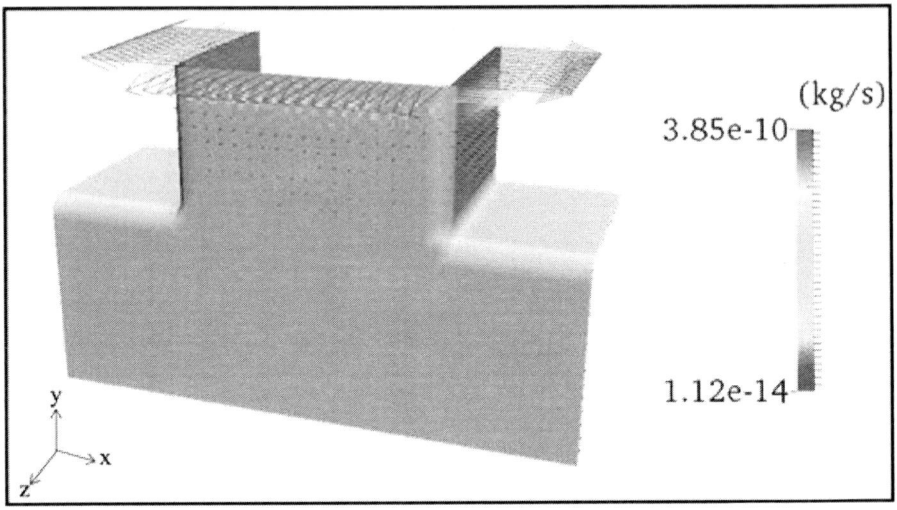

Figure 8. Mass flow rate on liquid-vapor interface viewed from rear.

between the casing and the wick, and the subcooling of the CC has been studied. Three-dimensional color renderings of pressure and temperature distributions were presented as well as 2D distributions of mass flow rates. Mass flow rates in the PTFE wick were more sensitive to the temperature distribution than those in the stainless steel wick. The evaporator heat-transfer coefficient for the stainless steel wick was approximately 10 times higher than that for the PTFE wick. On both wicks, heat fluxes concentrated at the boundary among the casing, the wick, and the grooves, and the result of the PTFE wick has larger distribution. The computational results showed that the length of the boundary per unit contact area is an important parameter in geometric optimization of the evaporator with respect to thermal performance. In our 3D analyses, the heat leak to the CC was estimated with high accuracy, and some cases showed small differences in the heat leaks between the PTFE and stainless steel wicks. In addition, the calculated results showed a heat-pipe effect in which evaporation and condensation occurred at the groove-wick interface. This effect was larger for the PTFE wick, which had a larger temperature distribution.

ACKNOWLEDGEMENTS

This research was partially supported by a Grant-in-Aid for JSPS Fellows (No. 25148) and by JST, PRESTO. A super computer system in the Information Technology center of Nagoya University was used for the calculations. LM and MP gratefully thank CNES and Airbus Defence and Space for their financial support.

REFERENCES

1. David, R., McGlen, R. and Kew, P. (2013) Heat Pipes: Theory, Design and Applications. 5th Edition, Butterworth- Heinemann, Oxford, 234-243.
2. Ku, J. (1999) Operating Characteristics of Loop Heat Pipes. SAE Technical Paper, No. 1999-01-2007.
3. Cao, Y.D. and Faghri, A. (1994) Analytical Solutions of Flow and Heat Transfer in a Porous Structure with Partial Heating and Evaporation on the Upper Surface. International Journal of Heat and Mass Transfer, 37, 1525-1533.
4. Cao, Y.D. and Faghri, A. (1994) Conjugate Analysis of a Flat-Plate Type Evaporator for Capillary Pumped Loops with Three-Dimensional Vapor Flow in the Groove. International Journal of Heat and Mass Transfer, 37, 401-409.
5. Ji, L. and Peterson, G.P. (2011) 3D Heat Transfer Analysis in a Loop Heat Pipe Evaporator with a Fully Saturated Wick. International Journal of Heat and Mass Transfer, 54, 564-574.
6. Li, Q., Zhao, K. and Xuan, Y.M. (2011) Simulation of Flow and Heat Transfer with Evaporation in a Porous Wick of a CPL Evaporator on Pore Scale by Lattice Boltzmann Method. International Journal of Heat and Mass Transfer, 54, 2890-2901.
7. Santos, P.H.D., Oliveira, A.A.M. and Bazzo, E. (2012) Analysis of the Onset of Drying of a Porous Ceramic Wick Applied to a Capillary Pumped Loop and a Loop Heat Pipe. Heat Pipe Science and Technology, an International Journal, 3, 125-151.
8. Kuroi, M. and Nagano, H. (2012) The Influence of Groove Shape on Loop Heat Pipe Performance. Heat Pipe Science and Technology, an International Journal, 3, 203-222.
9. Nishikawara, M., Nagano, H. and Prat, M. (2013) Evaporator Heat-Transfer Analysis of a Loop Heat Pipe with Low Thermal Conductivity Wicks. Proceedings of the 17th International Heat Pipe Conference, No. 18, Kanpur, 13-17 October 2013.

10. Kaya, T. and Goldak, J. (2006) Numerical Analysis of Heat and Mass Transfer in the Capillary Structure of a Loop Heat Pipe. International Journal of Heat and Mass Transfer, 49, 3211-3220.

11. Figus, C., Le Bray, Y., Bories, S. and Prat, M. (1999) Heat and Mass Transfer with Phase Change in a Porous Structure Partially Heated: Continuum Model and Pore Network Simulations. International Journal of Heat and Mass Transfer, 42, 2557-2569.

12. Prat, M. (2010) Pore Network Models for the Study of Transfers in the Porous Wick of Loop Heat Pipes. Heat Pipe Science and Technology, an International Journal, 1, 129-149.

13. Chernysheva, M.A. and Maydanik, Y.F. (2012) 3D-Model for Heat and Mass Transfer Simulation in Flat Evaporator of Copper-Water Loop Heat Pipe. Applied Thermal Engineering, 33-34, 124-134.

14. Nishikawara, M. and Nagano, H. (2012) Study on Evaporator Heat Transfer Characteristics of a Miniature Loop Heat Pipes with Low Thermal Conductivity Wicks. Proceedings of the 3rd International Forum on Heat Transfer, No. IFHT2012-013, Nagasaki, 13-15 November 2012.

15. Nagano, H., Fukuyoshi, F., Ogawa, H. and Nagai, H. (2011) Development of an Experimental Small Loop Heat Pipe with Polytetrafluoroethylene Wicks. Journal of Thermophysics and Heat Transfer, 25, 547-552.

16. Nishikawara, M. and Nagano, H. (2014) Parametric Experiments on a Miniature Loop Heat Pipe with PTFE Wicks. International Journal of Thermal Sciences, 85, 29-39.

17. Carey, V.P. (1992) Liquid-Vapor Phase-Change Phenomena. Taylor & Francis, London, 112.

18. Nishikawara, M., Nagano, H. and Kaya, T. (2013) Transient Thermo-Fluid Modeling of Loop Heat Pipes and Experimental Validation. Journal of Thermophysics and Heat Transfer, 27, 641-647.

19. Kaya, T., Pérez, R., Gregori, C. and Torres, A. (2008) Numerical Simulation of Transient Operation of Loop Heat Pipes. Applied Thermal Engineering, 28, 967-974.

20. Allen, T. (1997) Particle Size Measurement, Surface Area and Pore Size Determination. 5th Edition, Chapman and Hall, New York, 149.

21. Riehl, R.R. and Dos Santos, N. (2008) Loop Heat Pipe Performance Enhancement Using Primary Wick with Circumferential Grooves. Applied Thermal Engineering, 28, 1745-1755.

22. Maidanik, Y.F., Fershtater, Y.G. and Solodovnik, N.N. (1994) Design and Investigation of Methods of Regulation of Loop Heat Pipes for Terrestrial and Space Applications. Society of Automotive Engineering Technical Paper, No. 941407.

23. Kaya, T. and Hoang, T.T. (1999) Mathematical Modeling of Loop Heat Pipes and Experimental Validation. Journal of Thermophysics and Heat Transfer, 13, 314-320.

24. Singh, R., Akbarzadeh, A., Dixon, C. and Mochizuki, M. (2009) Theoretical Modeling of Miniature Loop Heat Pipe. Heat and Mass Transfer, 46, 209-224.

25. Bai, L.Z., Lin, G.P., Zhang, H.X. and Wen, D.S. (2009) Mathematical Modeling of Steady-State Operation of a Loop Heat Pipe. Applied Thermal Engineering, 29, 2643-2654.

26. Figus, C., Bories, S. and Prat, M. (1996) Investigation and Analysis of a Porous Evaporator for a Capillary Pumped Loop. Proceedings of the Engineering Systems and Design and Analysis Conference, Vol. 8, Montpellier, 1-4 July 1996, 99-106.

3
CHAPTER

PERFORMANCE OF SILOXANE MIXTURES IN A HIGH-TEMPERATURE ORGANIC RANKINE CYCLE CONSIDERING THE HEAT TRANSFER CHARACTERISTICS DURING EVAPORATION

Theresa Weith *, Florian Heberle, Markus Preißinger and Dieter Brüggemann

Lehrstuhl für Technische Thermodynamik und Transportprozesse, Zentrum für Energietechnik, Universität Bayreuth, Universitätsstraße 30, 95447 Bayreuth, Germany

ABSTRACT

The application of the Organic Rankine Cycle to high temperature heat sources is investigated on the case study of waste heat recovery from a selected biogas plant. Two different modes of operation are distinguished: pure electric power and combined heat and power generation. The siloxanes hexamethyldisiloxane (MM) and octamethyltrisiloxane (MDM) are chosen as working fluids. Moreover, the effect of using mixtures of these components is analysed. Regarding pure electricity generation, process simulations using the simulation tool Aspen Plus show an increase in second law efficiency of 1.3% in case of 97/03 wt % MM/MDM-mixture, whereas for the combined heat and power mode a 60/40 wt % MM/MDM-mixture yields the highest efficiency with an increase of nearly 3% compared to most efficient pure fluid. Next to thermodynamic analysis, measurements of heat transfer coefficients of these siloxanes as well as their mixtures are conducted and Kandlikar's correlation is chosen to describe the results. Based on that, heat exchanger areas for preheater and evaporator are calculated in order to check whether the poorer heat transfer characteristics of mixtures devalue their efficiency benefit due to increased heat transfer areas. Results show higher heat transfer areas of 0.9% and 14%, respectively, compared to MM.

Keywords: Organic Rankine Cycle; siloxanes; high temperature heat source; zeotropic mixtures; heat transfer coefficients; heat exchange area

1. INTRODUCTION

The use of zeotropic mixtures in Organic Rankine Cycle (ORC) applications has been widely investigated during the last years. The focus of this research is mainly on low temperature applications like geothermal systems [1,2,3,4]. In these cases an efficiency increase of up to 20% is possible by using mixtures compared to the most efficient pure component due to a better glide matching during condensation and evaporation caused by non-isothermal phase change of zeotropic mixtures. Recent investigations also figure the combination of zeotropic mixtures and a transcritical mode of operation due to a better matching of temperature profiles [5,6]. Next to the non-isothermal phase change during condensation resulting from the use of zeotropic mixtures, the evaporation phase change also occurs non-isothermally, due to the supercritical mode of operation. This results in an increase in second law efficiency of up to 60% [5]. Beside geothermal power generation, the benefit of using zeotropic mixtures is also investigated for other specific applications, like for example in the field of heat recovery from engines [7,8]. Furthermore, the first studies that focus on zeotropic mixtures for high temperature ORC have been published. Along with Angelino and Colonna [9,10], who investigated multi-component mixtures of linear siloxanes, Dong et al. [11] show results of the performance of hexamethyldisiloxane (MM)/octamethyltrisiloxane (MDM)-mixtures at a heat source temperature of 280 °C. Compared to low temperature applications, the efficiency increase achieved with mixtures in high temperature ORCs is less. Chys et al. [12] describe a benefit of nearly

16% for a heat source temperature of 150 °C, whereas at 250 °C the increase is less than 6%.

Next to this advantage of zeotropic mixtures with respect to cycle efficiency, there is also an inherent drawback of using them in ORC-applications caused by their generally reduced heat transfer characteristics compared to pure fluids [13,14]. This fact is often neglected when analysing mixtures in ORC-systems, but regarding sizing of heat exchangers the knowledge of this degradation in heat transfer coefficient is important. Looking at common refrigerants, measurements of heat transfer coefficients of those as well as their mixtures are already available in literature as their long-serving use in refrigeration-systems has driven this research [15,16,17,18,19,20]. However, in case of siloxane mixtures as well as pure siloxanes up to now there has been a lack of data.

The present study shows a thermodynamic analysis of a high temperature ORC using the linear siloxanes hexamethyldisiloxane (MM) and octamethyltrisiloxane (MDM) as well as their mixtures. Moreover, the results of measurements of heat transfer coefficients of these fluids and their mixtures in steps of 20 wt % MM are presented. Kandlikar's correlation is adapted to the experimental values and is then used for calculating the required heat exchanger area for preheating and evaporation.

2. THERMODYNAMIC PROCESS ANALYSIS

In this study the recovery of waste heat from biogas plants is chosen exemplarily as one application of the ORC. The specifications of the exhaust gas from the biogas-engine (MAN E 2842 LE 322, MAN Truck & Bus AG, Nuremberg, Germany) of the selected biogas plant are summarized in Table 1.

Table 1. Parameters of heat source (EG).

$T_{EG\text{-}IN}$ (K)	$\dot{m}_{EG}$ (kg/s)	p_{EG} (bar)	Composition (mol %)				
			N_2	H_2O	CO_2	O_2	Ar
733.15	0.48	1	64.55	18.15	9.74	6.63	0.93

Siloxanes were chosen as a working fluid due to their good performance in high temperature applications [9,21,22]. In particular, the two linear siloxanes hexamethyldisiloxane (MM) and octamethyltrisiloxane (MDM) as well as their mixtures are investigated. First degradation measurements of

MM show its thermal stability up to a temperature of 300 °C [23]. Therefore, with regard to the high heat source temperature, an additional intermediate circuit is considered in the simulations in order to prevent the working fluid from decomposition. Dowtherm G is chosen as thermal oil due to its high thermal stability [24].

2.1. Methodology

Simulations have been performed using the simulation tool Aspen Plus [25]. As property method the PENG-ROB (Peng-Robinson equation of state) model is used for the exhaust gas, air, and Dowtherm G as well as the ORC-fluid. In the case of water, STEAM-NBS (NBS/NRC Steam Tables) is applied [26]. Figure 1 shows the flowsheet of the process. It consists of the ORC, in which an internal recuperator is applied for the purpose of efficiency improvement, and the additional thermal oil loop. All heat exchangers are operating in counter flow mode and pressure drops are neglected.

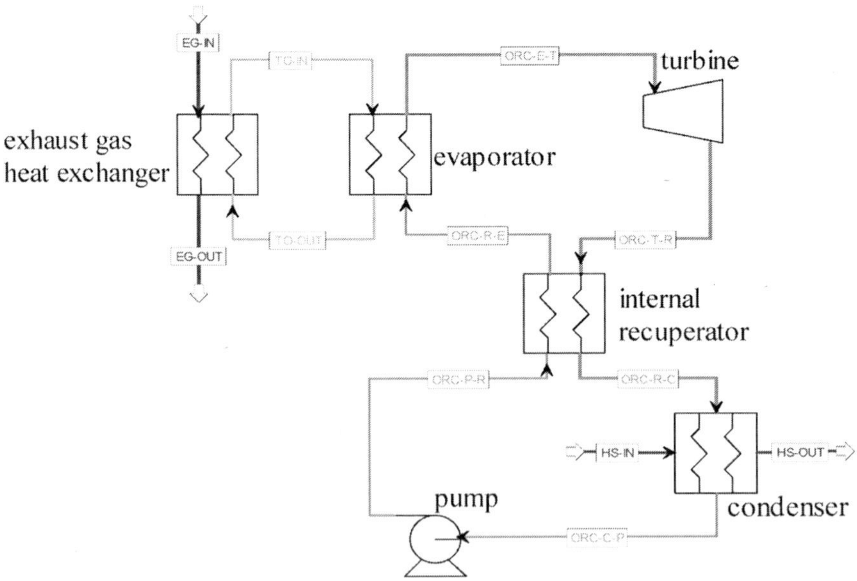

Figure 1. Aspen Plus flowsheet.

For the heat sink, two different scenarios are analysed. The first one represents pure electricity generation out of waste heat assuming direct air cooled condensation, whereas the second one considers an additional use of the condenser duty in a heating network using water. Table 2 lists the general parameters of the heat sink for those two cases.

Table 2. Heat sink (HS) parameters.

Case	Fluid	$T_{HS\text{-}IN}$ (K)	$T_{HS\text{-}OUT}$ (K)	$\dot{m}_{HS}$ (kg/s)	p_{HS} (bar)
heat sink (electricity)	air	288.15	303.15	variable	1
heat sink (CHP)	water	323.15	343.15	variable	2

As boundary conditions, the pinch point differences in the heat exchangers are kept constant (Table 3). Only the pinch point differences in the condenser and internal recuperator can reach values higher than those given in Table 3. In the condenser this occurs due to the limitation of the lower working pressure to a minimum value of 0.1 bar. Regarding fluid mixtures, partial condensation can take place in the internal recuperator. If this occurs, the boundary condition is switched. Instead of the fixed pinch point temperature difference a vapour fraction of one is set for the hot outlet stream of the recuperator (ORC-R-C). As a result, pinch point differences can reach higher values. Due to the heat source parameters, the investigated systems are small-scale ORCs with an electrical power output in the kilowatt range. Therefore, the isentropic efficiency of the turbine is set to the relatively small value of 60%.

Table 3. Boundary conditions.

$PP_{\text{EG heat exchanger}}$ (K)	$PP_{\text{evaporator}}$ (K)	$PP_{\text{condenser}}$ (K) (min)	$PP_{\text{internal recuoerator}}$ (K) (min)	η_{turbine} (%)	$\eta_{\text{generator}}$ (%)	η_{pump} (%)	η_{driver} (%)
20	30	10	10	60	90	70	90

To compare the simulation results of the different fluids, the exergetic efficiency is chosen as a criterion. For pure generation of electric energy, the exergetic efficiency is defined by the net power P_{net} (power of turbine $P_{el,turb}$ minus power of pump P_{pump}) divided by the exergy of heat source $\dot{E}_{EG}$:

$$\eta_{ex,electricity} = \frac{P_{\text{net}}}{\dot{E}_{\text{EG}}} = \frac{P_{\text{el,turb}} - P_{\text{pump}}}{\dot{m}_{\text{EG}}\left[h_{\text{EG,in}} - h_0 - T_0\left(s_{\text{EG,in}} - s_0\right)\right]} \quad (1)$$

For calculating $\dot{E}_{EG}$, mass flow rate $\dot{m}_{EG}$, enthalpy $h_{EG,in}$ as well as entropy of heat source $s_{EG,in}$ are used. The index 0 corresponds to the reference state

(288.15 K, 1 bar). In the case of combined heat and power generation the exergy content of the condenser duty $\dot{E}_{cond}$ has to be taken into account:

$$\eta_{ex,\,CHP} = \frac{P_{net} + \dot{E}_{cond}}{\dot{E}_{EG}} \tag{2}$$

$$\dot{E}_{cond} = \dot{m}_{HS}\left[h_{HS,out} - h_{HS,in} - T_0\left(s_{HS,out} - s_{HS,in}\right)\right] \tag{3}$$

It is calculated by using inlet and outlet enthalpies ($h_{HS,in/out}$) and entropies ($s_{HS,in/out}$) of the cooling water as well as its mass flow rate $\dot{m}_{HS}$.

For evaluating the performance during evaporation and condensation, the irreversibilities of the evaporator and condenser are calculated by using respective inlet and outlet enthalpies and entropies:

$$\dot{I}_{evap} = T_0 \dot{m}_{ORC}\left[s_{ORC\text{-}E\text{-}T} - s_{ORC\text{-}R\text{-}E} - \frac{h_{ORC\text{-}E\text{-}T} - h_{ORC\text{-}R\text{-}E}}{T_{m,evap}}\right] \tag{4}$$

$$\dot{I}_{cond} = T_0 \dot{m}_{ORC}\left[s_{ORC\text{-}C\text{-}P} - s_{ORC\text{-}R\text{-}C} - \frac{h_{ORC\text{-}C\text{-}P} - h_{ORC\text{-}R\text{-}C}}{T_{m,cond}}\right] \tag{5}$$

The logarithmic mean temperatures are defined as follows:

$$T_m = \frac{T_{in} - T_{out}}{\ln\left(\frac{T_{in}}{T_{out}}\right)} \tag{6}$$

In case of the condenser T_{in} and T_{out} describe the inlet and outlet temperatures of the heat sink. For the irreversibilities of the evaporator the temperatures of Dowtherm G have to be used.

2.2. Results of Process Simulation

Due to the available heat source, the applied ORC shows power outputs in the low power range (P_{net} < 30 kW). The results of the process simulation are displayed in Figure 2. On the left hand side the exergetic efficiency against the mass fraction of MM in case of pure electricity generation is presented, whereas the results for the combined heat and power case are shown on the right hand side. Next to the efficiency, the corresponding upper working pressure, that yields the highest efficiency in each case, is added in the diagrams.

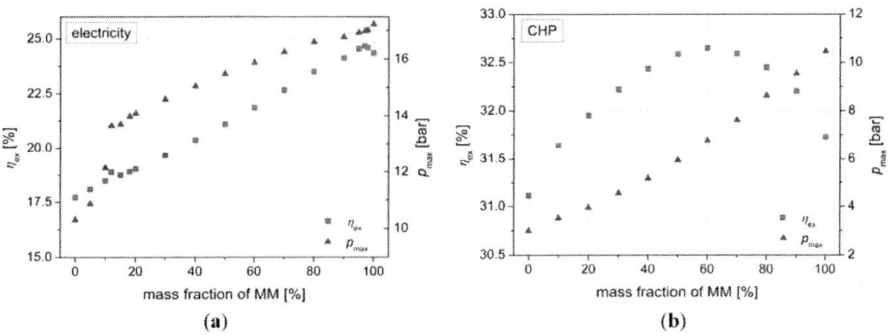

Figure 2. Exergetic efficiencies and respective maximum working pressures against mass fraction of MM ((**a**): pure electricity generation, (**b**): combined heat and power generation).

Regarding the case of pure electricity generation and using steps of mass fraction of MM of 10%, it seems that there is a nearly linear behaviour of exergetic efficiency with respect to mass fraction of MM. A closer look reveals that at high mass fractions of MM there might be a deviation from this trend. Therefore, additional steps of 95%, 97% and 98% MM are added and an efficiency increase of nearly 1.3% compared to the most efficient pure component, MM, occurs for 97% MM. A similar effect between mass fractions of 10% and 20% occur, but efficiencies do not exceed that of MM.

A quite different trend with respect to mass fraction of MM can be seen for the combined heat and power production. Compared to the pure fluid with the best exergetic efficiency (MM: η_{ex} = 31.73%) an efficiency increase of 2.9% is possible when using mixtures (60 wt % MM: η_{ex} = 32.65%). Next to this improved cycle efficiency, lower working pressures are needed compared to pure MM. As shown in Figure 3, this increase in second law efficiency derives from an irreversibility minimum in the condenser. Regarding the irreversibilities in the evaporator, smaller differences between the fluids can be observed. Therefore, they do not affect the overall cycle efficiency.

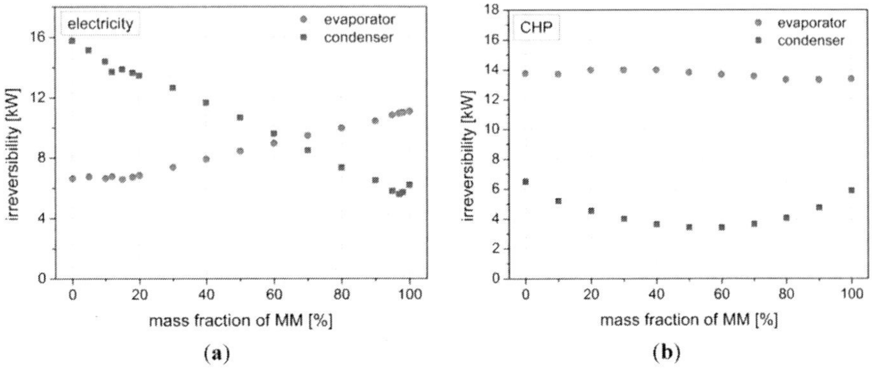

Figure 3. Irreversibilities in evaporator and condenser in case of (**a**) pure electricity generation and (**b**) combined heat and power generation.

The reason for this behaviour can be revealed by regarding the temperature profiles during evaporation and condensation (Figure 4). It can be seen that the temperature glide during evaporation is negligible compared to that during condensation. Therefore, as also published by other authors [3], a good glide matching during condensation, which results in a lower average heat rejection temperature of the cycle, is decisive for cycle efficiency. Figure 4b fairly shows the accordance in the slope of the temperature profile of heat sink and the most efficient mixture.

All in all, the results show that an increase in efficiency of nearly 1.3% (electricity) or 3% (CHP) is possible by using mixtures of MM and MDM, but, whether a use of mixtures is really advantageous or not depends, next to the pure thermodynamic performance, on several additional aspects. In this regard, the effect of an eventual composition shift on the cycle efficiency has to be considered. Moreover, the influence of mixtures on the performance of cycle components (e.g., the respective turbine efficiency) has to be evaluated. As heat transfer coefficients of mixtures are said to be lower than those of the pure components, it is crucial to investigate the required size of heat exchangers. In order to estimate this heat transfer area, proper correlations to describe the heat transfer characteristics are necessary. To identify them, measurements of heat transfer coefficients have to be conducted. The main parameter for a waste heat recovery system is the coupling with the heat source and, therefore, the evaporation process. Hence, this study focuses in a first step on evaporation heat transfer.

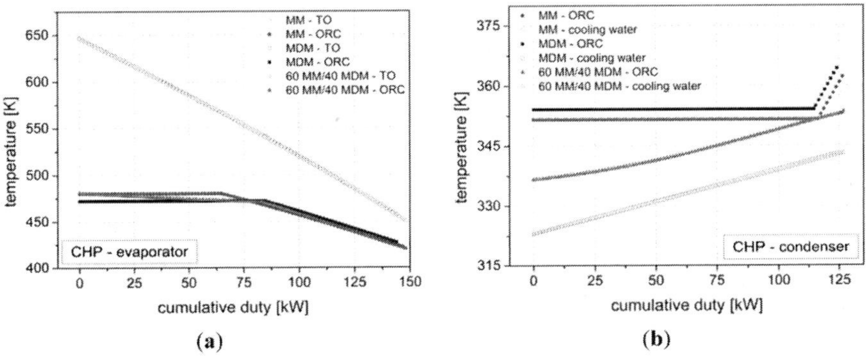

Figure 4. Temperature profiles during (**a**) evaporation and (**b**) condensation for the pure fluids as well as for the most efficient mixture (60 wt % MM) in case of CHP.

3. HEAT TRANSFER COEFFICIENTS OF SILOXANES AND SILOXANE MIXTURES

3.1. Experimental Setup

To measure the flow boiling heat transfer coefficients of the investigated siloxanes MM and MDM as well as their mixtures, a test rig (Figure 5) was built consisting of a pump, which raises the working fluid to the desired pressure, an electrically heated flow heater, which preheats the siloxane up to its saturation temperature, a test section, in which the fluid is evaporated, two expansions devices to reduce the pressure and a double-pipe heat exchanger to condense the fluid.

All components and pipes are thermally isolated to the environment by a glass wool layer. The main test section (Figure 6) consists in a smooth, horizontal tube with one U-turn, which is electrically heated by Joule effect. It is electrically isolated from the rest of the test rig by a high-temperature plastic.

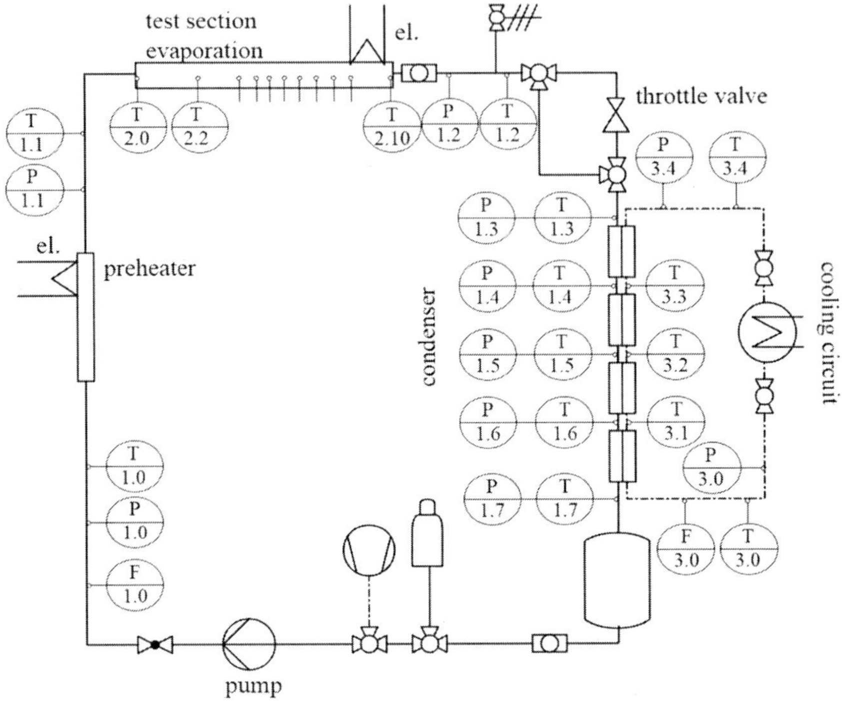

Figure 5. Test rig for measuring heat transfer coefficients.

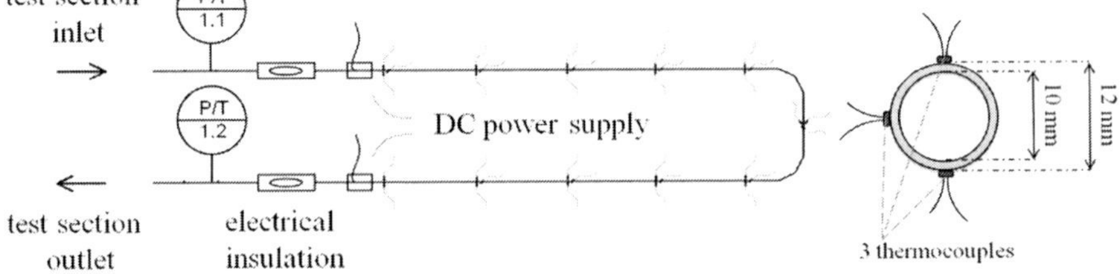

Figure 6. Test section for flow boiling.

Temperature and pressure is measured at the inlet as well as the outlet of the evaporator. Eleven measurement sites are placed alongside the evaporator in order to investigate heat transfer over the whole test section. At each measurement site three thermocouples are mounted at the tube.

The overall outer wall temperature $T_{w,o}$ for one measurement site is therefore calculated by:

$$T_{w,o} = \frac{T_{w,t} + 2 \cdot T_{w,m} + T_{w,b}}{4} \quad (7)$$

using the temperatures at the top ($T_{w,t}$), middle ($T_{w,m}$) and bottom ($T_{w,b}$) of the tube.

The local heat transfer coefficient of evaporation h_{evap} is evaluated by:

$$h_{evap} = \frac{\dot{q}}{T_{w,i} - T_{sat}(x,p)} = \frac{\frac{P_{el}}{A}}{T_{w,i} - T_{sat}(x,p)} \quad (8)$$

Thereby, P_{el} corresponds to the power supplied by the DC power device. The temperature at the inner face of the tube, $T_{w,i}$, can be determined with respect to the outer wall temperature, $T_{w,o}$, by applying the law of heat conduction. T_{sat} represents the bulk temperature and therefore corresponds to the respective boiling point temperature, which depends on pressure and regarding mixtures as well on vapour quality. Its value is obtained by REFPROP (Reference Fluid Thermodynamic and Transport Properties Database) [27] assuming a linear pressure profile along the test section.

3.2. Experimental Results

Experiments have been performed at constant critical pressure as well as constant saturation temperature. Mixtures in steps of 20 wt % MM have been investigated. Figure 7 shows the results of the average heat transfer coefficients during evaporation with respect to mass fraction of MM. Respective uncertainties in measurements are below 30% [28].

It is obvious that the heat transfer coefficients of the siloxane mixtures deviate from the theoretical ideal values. The ideal curve is just a linear interpolation between the heat transfer coefficients of the pure components. In fact, it does not have any real physical meaning, but it can be used to describe the effect that the heat transfer coefficients of mixtures can be lower than those of the pure components. In case of constant reduced pressure, the maximum deviation of 46% occurs at 20% MM. The heat transfer coefficient at constant saturation temperature deviates maximum 42% from the ideal value, again at 20% MM.

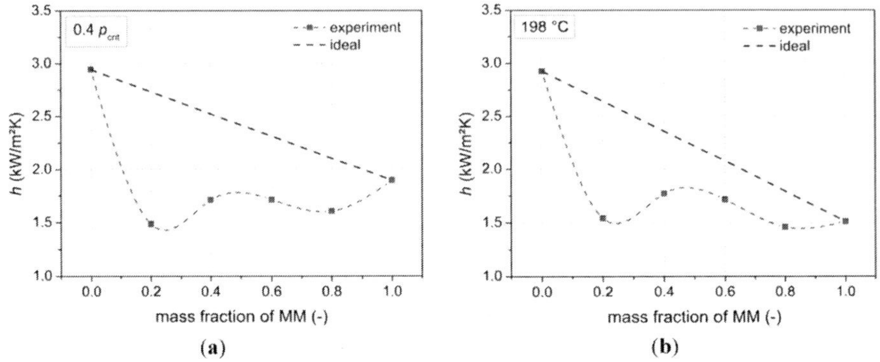

Figure 7. Average heat transfer coefficients for different mass fractions of MM at a pressure of (**a**) 0.4 p_{crit} [28] as well as (**b**) a saturation temperature of 198 °C.

3.3. Examination of Appropriate Correlations

A lot of different correlations for flow boiling of pure fluids as well as mixtures have been proposed during the last decades. Nevertheless, none of those correlations are based on high-temperature fluids like the investigated siloxanes. Due to that, as a basis, a correlation has to be chosen that offers a fluid-specific parameter in order to adjust it to the experimental data. Such correlations are given by Kandlikar [29] for pure fluids as well as mixtures. The investigation of the adaption of Kandlikar's correlations to the experiments with siloxanes and their mixtures revealed that they are able to represent the general trend of heat transfer coefficient with respect to vapour quality and mass flux density quite well [28]. Therefore, the present study focuses on this correlation.

Regarding flow boiling in plain tubes, Kandlikar [29] suggests the following expression for the heat transfer coefficient of pure fluids (α_{TP}):

$$\alpha_{TP} = \max\left(\alpha_{NBD}, \alpha_{CBD}\right) \tag{9}$$

where α_{NBD} describes the heat transfer coefficient in the nucleate boiling region:

$$\alpha_{NBD} = 0.6683\ Co^{-0.2}(1-x)^{0.8}\ \alpha_{lo} + 1058.0\ Bo^{0.7}(1-x)^{0.8}F_{Fl}\alpha_{lo} \tag{10}$$

whereas α_{CBD} is used for regions where convective boiling is dominant:

$$\alpha_{CBD} = 1.136\,Co^{-0.9}(1-x)^{0.8}\,\alpha_{lo} + 667.2\,Bo^{0.7}(1-x)^{0.8}F_{Fl}\alpha_{lo} \quad (11)$$

In both cases the heat transfer coefficient with respect to vapour quality x depends on the convection number Co, the boiling number Bo as well as on a fluid-surface parameter F_{Fl}. For the single-phase heat transfer coefficient α_{lo} the correlations by Petukhov and Popov (Equation (12), for $0.5 \leq Pr_L \leq 2000$ and $10^4 \leq Re_{LO} \leq 5 \times 10^6$) or Gnielinski (Equation (13), for $0.5 \leq Pr_L \leq 2000$ and $2300 \leq Re_{LO} \leq 10^4$) are applied:

$$\alpha_{lo,PaP} = \frac{\lambda_l}{d_i}\frac{Re_{lo}\,Pr_L\,(f/2)}{1.07 + 12.7\left(Pr_L^{2/3} - 1\right)(f/2)^{0.5}} \quad (12)$$

$$\alpha_{lo,G} = \frac{\lambda_l}{d_i}\frac{(Re_{lo} - 1000)\,Pr_L\,(f/2)}{1.07 + 12.7\left(Pr_L^{2/3} - 1\right)(f/2)^{0.5}} \quad (13)$$

$$f = [1.58\ln(Re_{lo}) - 3.28]^{-2} \quad (14)$$

These correlations are based on the Reynolds number Re, the Prandtl number Pr and the friction factor f. In addition, the liquid thermal conductivity λ_l as well as the inner diameter d_i of the tube contribute to the heat transfer coefficient.

In case of fluid mixtures, Kandlikar distinguishes between three different regions according to the volatility parameter V_1 and the boiling number Bo [29]:

1. Near azeotropic region—$V_1 < 0.03$. In this domain the correlation for pure fluids is adopted (Equations (9)–(14)).
2. Moderate diffusion-induced suppression region—$0.03 < V_1 < 0.2$ and $Bo > 10^{-4}$. Heat transfer is dominated by convection. Therefore, Equation (12) is used to predict heat transfer coefficients.
3. Severe diffusion-induced suppression region—$0.03 < V_1 < 0.2$ and $Bo \leq 10^{-4}$ or $V_1 \geq 0.2$. This region is still dominated by convection.

Moreover, additional mass diffusion resistance due to large composition differences has to be taken into account. Due to that, Equation (12) is extended by the diffusion-induced suppression factor F_D:

$$\alpha_{CBD} = 1.136\,\mathrm{Co}^{-0.9}(1-x)^{0.8}\,\alpha_{lo} + 667.2\,\mathrm{Bo}^{0.7}(1-x)^{0.8}F_{Fl}\alpha_{lo}F_D \quad (15)$$

$$F_D = \frac{0.678}{1+V_1} \quad (16)$$

where the volatility parameter V_1 is defined by:

$$V_1 = \frac{c_{p,l}}{\Delta h_{LG}}\left(\frac{\kappa}{D_{12}}\right)^{0.5}\left|(y_1 - x_1)\left(\frac{dT}{dx_1}\right)\right| \quad (17)$$

using the liquid specific heat $c_{p,l}$, the latent heat of vaporisation Δh_{LG}, the thermal diffusivity κ and the diffusion coefficient D_{12}. Moreover, the mass fraction of component 1 in vapour (y_1) and liquid (x_1) phase as well as the slope of the bubble point curve (dT/dx_1) contribute.

At first, Kandlikar's correlation was applied to the experimental test conditions while using F_{Fl} = 1 as suggested by Kandlikar for all fluids in stainless steel tubes. The results are demonstrated in Figure 8.

It can be seen that the absolute values between the correlation and the measurements differ, especially for the results of MM. For constant reduced pressure, the value of the experimental results of MM lies 46% below those of the correlation. In case of constant evaporation temperature the deviation is even 58%.

In order to obtain better agreement between the experimental and the theoretical heat transfer coefficients, the fluid-surface parameter F_{Fl} is used to adapt the Using $F_{Fl,MM}$ = 0.1 and $F_{Fl,MDM}$ = 0.8 results in maximum deviations of 15% in case of MM and 1.5% regarding MDM. For mixtures, the fluid-surface parameter is derived from those of the pure components as suggested by Kandlikar:

$$F_{Fl,\,mixture} = x_{MM}F_{Fl,MM} + x_{MDM}F_{Fl,MDM} \quad (18)$$

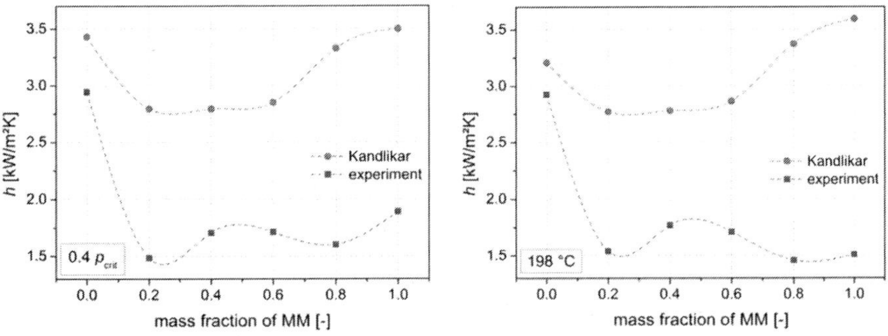

(a) (b)

Figure 8. Comparison of experimental results and Kandlikar's correlation (F_{Fl} = 1): Average heat transfer coefficients for different mass fractions of MM at a pressure of (**a**) 0.4 p_{crit} as well as (**b**) a saturation temperature of 198 °C.

Hereby, x_{MM} corresponds to the mass fraction of MM in liquid phase, x_{MDM} to that of MDM, respectively.

The results of applying these parameter values in the correlation are presented in Figure 9. The degradation in heat transfer coefficient in case of mixtures is small compared to the experimental data. In both cases, the mean deviations of the experimental data to the correlation are below 17%. Maximum deviations appear at 20% MM with a value of nearly −37%.

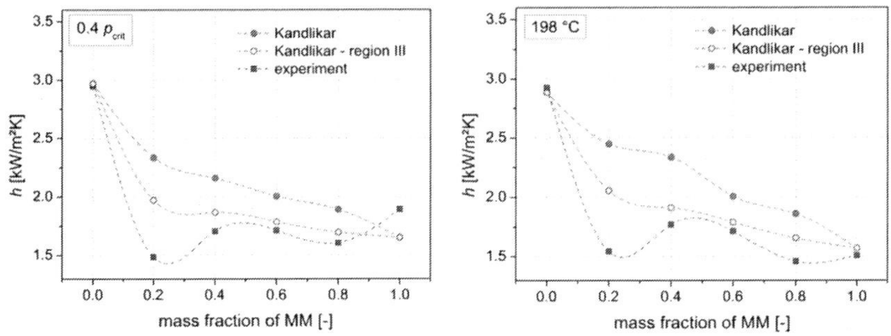

(a) (b)

Figure 9. Comparison of experimental results and Kandlikar's correlation ($F_{Fl,MM}$ = 0.1, $F_{Fl,MDM}$ = 0.8): Average heat transfer coefficients for different mass fractions of MM at a pressure of (**a**) 0.4 p_{crit} as well as (**b**) a saturation temperature of 198 °C.

Due to the investigated boundary conditions, only region I and II from Kandlikar's correlation are applied. As the experimental data still deviate from the adjusted correlation, the results by using the severe diffusion-induced suppression region (region III) (even though the volatility parameter or the Bo-Number would suggest another region) are analysed, too. The respective average heat transfer coefficients are added in Figure 9. By using this equation, maximum deviations of solely 25% are present. Mean deviations of the experimental data to the correlation are below 10%. Therefore, it is recommended to use the correlation given in region III together with the adjusted values of F_{Fl} when calculating heat transfer coefficients of MM/MDM-mixtures.

4. ESTIMATION OF REQUIRED HEAT EXCHANGER AREA

In order to reveal, whether the reduced heat transfer characteristics of fluid mixtures compensate their slight predominance in case of system efficiency, required heat exchange areas are investigated. For that purpose, a shell-and-tube heat exchanger with one shell pass and two tube passes is used as preheater and evaporator. Table 4 lists the main specifications and assumptions.

Table 4. Specifications and assumptions for the shell and tube heat exchanger.

Geometry Data			Maximum Velocities (m/s) [30,31]		
outer tube radius r_a (mm)	wall thickness Δd (mm)	thermal conductivity λ_{steel} (W/mK)	ORC liquid	ORC vapour	Dowtherm
9.5	2.1	15	4	20	4

The general approach for calculating the heat exchanger area is derived from Roetzel and Spang [32]. It is based on the Number of Transfer Units (NTU)-method. The NTU of the real shell and tube heat exchanger is obtained by applying a correction factor to the NTU of an ideal counterflow heat exchanger. For calculating the heat transfer coefficients at the inner and outer tube different models are applied. For Dowtherm G as well as for the ORC-fluid in the preheater, the correlation for turbulent flow by Gnielinski [33] is chosen, whereas for the evaporation of the ORC-fluid Kandlikar's correlation, as presented in Section 3, is used. The whole calculation is performed in Matlab [34]. Fluid properties are derived from REFPROP [27].

The calculations of required heat transfer area are performed for the pure fluids as well as for one more efficient mixture for the case of pure electricity and combined heat and power generation, respectively. Table 5 and Table 6 list the respective process parameters that derived from the thermodynamic analysis in Section 2. These values represent the input parameters of the heat exchanger calculations.

Table 5. Process parameters for pure electric energy generation.

Fluid	$\dot{m}_{TO}$ (kg/s)	$T_{TO\text{-}IN}$ (K)	$T_{TO\text{-}OUT}$ (K)	$\dot{m}_{ORC}$ (kg/s)	$T_{ORC\text{-}R\text{-}E}$ (K)	$T_{ORC\text{-}E\text{-}T}$ (K)	p_{ORC} (bar)
MM	0.35	645.94	449.68	0.50	419.68	511.59	17.23
MDM	0.35	645.41	502.74	0.51	472.74	541.93	10.30
MM/MDM 95/05	0.35	646.32	452.51	0.50	422.51	513.32	16.92

Table 6. Process parameters for combined heat and power generation.

Fluid	$\dot{m}_{TO}$ (kg/s)	$T_{TO\text{-}IN}$ (K)	$T_{TO\text{-}OUT}$ (K)	$\dot{m}_{ORC}$ (kg/s)	$T_{ORC\text{-}R\text{-}E}$ (K)	$T_{ORC\text{-}E\text{-}T}$ (K)	p_{ORC} (bar)
MM	0.35	646.32	452.58	0.57	422.56	480.22	10.46
MDM	0.35	647.04	457.78	0.65	427.78	472.37	3.00
MM/MDM 60/40	0.35	645.99	450.09	0.57	420.30	480.90	6.77

The results of the heat exchanger analysis are summarised in Table 7 and Table 8. In both cases MM requires the smallest heat transfer area. For pure electricity generation MDM shows the biggest heat exchanger areas. Regarding the evaporator (EVAP), 19% more heat compared to MM has to be transferred, but the heat transfer coefficient of evaporation exceeds that of MM by 53%. Due to that, an additional aspect has to contribute to the increased heat transfer area of 56%. Table 5 shows that mass flow rates of all three fluids are in the same range, but MDM shows the lowest pressure as well as temperature differences. Therefore, process parameters like temperatures and the respective fluid properties affect the required area just as well as pure heat transfer characteristics of fluids.

Table 7. Results of heat exchanger calculations for the case of pure electric power generation.

Pure Electricity	Heat Exchange Area (m^2)		Number of Tubes		Heat Transfer Coefficient ($kW/m^2 \cdot K$) ORC/Dowtherm		Heat Duty (kW)	
	PH	*EVAP*	*PH*	*EVAP*	*PH*	*EVAP*	*PH*	*EVAP*
MM	0.725	0.120	2	2	5.761/7.202	26.414/8.343	113.35	30.725
MDM	0.741	0.187	2	2	5.141/7.450	40.502/8.256	86.901	36.500
MM/MDM 95/05	0.728	0.125	2	2	5.699/7.201	32.575/8.336	111.6	32.135

Table 8. Results of heat exchanger calculations for the case of combined heat and power generation.

CHP	Heat Exchange Area (m^2)		Number of Tubes		Heat Transfer Coefficient ($kW/m^2 \cdot K$) ORC/Dowtherm		Heat Duty (kW)	
	PH	*EVAP*	*PH*	*EVAP*	*PH*	*EVAP*	*PH*	*EVAP*
MM	0.583	0.270	2	3	5.756/6.942	34.394/8.660	77.76	65.276
MDM	0.536	0.379	2	9	4.869/6.806	60.753/10.312	63.873	83.795
MM/MDM 60/40	0.602	0.373	2	4	5.245/6.762	37.859/8.914	64.009	80.192

In case of the mixture the evaporator needs a 4.2% enlarged area compared to that of MM. Process parameters are quite similar. The heat transfer coefficient of the mixture lies 23.3% above that of MM, whereas a higher heat duty of 4.6% is transferred. Together with slightly reduced temperature differences a higher area is required. Regarding the whole process of heat input using the mixture, 0.9% higher heat transfer areas result compared to MM.

For the case of the combined heat and power generation similar trends occur. Regarding the evaporator, MDM requires a 40.4% higher area than MM. Heat transfer coefficient is 76.6%, heat duty 28.4% above the values of MM. A 14% higher mass flow rate as well as again lower pressure and temperature differences affect the required area, too. As the number of tubes is set according to the maximum velocity in the tubes, more tubes are required in the evaporator for MDM due to higher mass flow rates. As the system pressure in case of combined heat and power generation is lower than in the case of pure electricity generation, a higher part of the area is omitted to the evaporator due to increased heat of vaporisation.

The process parameters in case of the mixture again correspond more to those of MM. The heat duty in the evaporator exceeds that of MM by 22.9%. Heat transfer coefficients are 10.1% higher. An increase in heat transfer area of 38.15% is required. Again, compared to MM, slightly lower temperature differences and pressures are present. The total heat transfer area of preheater (PH) and evaporator exceeds that of MM by 14%.

Summing up it has to be stated that the pure heat transfer characteristics of fluids do not necessarily determine the size of the heat exchanger area. Process parameters play an even comparable role. For the investigated cases, the most efficient mixtures require an increase in heat transfer area for the heat input compared to the most efficient pure component of 0.9% and 14%, respectively. This corresponds to an increase in second law efficiency of 0.8% and 2.9%, respectively. As stated in the introduction, the potential of efficiency increase in high temperature ORC is low compared to low temperature applications. Heberle [35] reports an efficiency increase of almost 16% for a geothermal application using mixtures of R245fa and R365mfc. But the required heat transfer area (evaporator + preheater) for the most efficient mixture composition exceeds that of the most efficient pure fluid by 83% [35]. This effect is strongly influenced by the highly reduced heat transfer coefficients of R245fa/R365mfc-mixture compared to their pure components. Reductions compared to the pure fluids of up to nearly 50% are observed [36], whereas for the here investigated siloxane mixtures maximum deviations from the pure component with the lowest heat transfer coefficient are −21.5%. Therefore, it can be concluded, that even though comparably low efficiency increases are possible in high temperature applications, the required additional heat transfer area for mixtures is reduced compared to low temperature systems.

As the present investigations use an additional thermal oil loop, liquid heat transfer coefficients of heat source and ORC-fluid are quite similar. Therefore, the heat transfer coefficient of the ORC-fluid during evaporation can have a strong effect on the required heat transfer area. Regarding applications with direct use of exhaust gas, this effect is diminished or even negligible due to the generally low heat transfer characteristics of the exhaust gas. In these cases, already slight increases in efficiency by using mixtures can justify their application.

Next to the evaporation process, it is necessary to investigate the heat transfer areas of internal heat exchanger and condenser as well. As during internal heat exchange no phase change occurs, degradation of heat transfer coefficients of mixtures are not expected and, therefore, there should be no additional increase in heat transfer area. In contrary, during condensation, reduced heat transfer coefficients in case of mixtures are obvious. Thus, the heat transfer area of the condenser has to be calculated, too. In case of directly air cooled processes, the effect of heat transfer degradation of mixtures is again negligible due to the worse heat transfer characteristics of air. But, regarding the case of combined heat and power generation, where water is used as a cooling medium, the effect of the lower heat transfer coefficients of mixtures can again result in larger heat exchange areas.

5. CONCLUSIONS

The performance of the two linear siloxanes MM and MDM as well as their mixtures in a high temperature ORC are analysed using the example of waste heat from biogas plant. In case of pure electricity generation only slight improvements (1.3%) in second law efficiency can be achieved by using a mixture of 97 wt % MM. An efficiency increase of nearly 3% is obtained for the combined heat and power generation with the mixture 60/40-wt % MM/MDM. As the potential of efficiency increase drops with rising heat source temperature, these results seem reasonable compared to those given in literature.

Next to the thermodynamic analysis, measurements of heat transfer coefficients of the pure siloxanes as well as their mixtures (steps of 20 wt % MM) have been performed. Results show degradations of heat transfer coefficients compared to the ideal values of up to 46%. Kandlikar's correlation is used to describe the experimental data. In order to achieve a good agreement with the real values, the fluid-surface parameter was used to adjust the correlation. Moreover, the severe-diffusion induced region was chosen. By using this adapted correlation, mean deviations of less than 10% between the experimental data and the correlation are present.

Based on all these results, required heat transfer areas of evaporator and preheater are estimated assuming shell-and-tube heat exchangers. Calculations are performed for the case of pure electricity generation as well as combined heat and power generation for the pure components and one mixture. Results show an increased heat transfer area in case of the mixtures of around 1% (pure electricity, 95 wt % MM, efficiency increase: ≈ 1%) and 14% (combined heat and power, 60 wt % MM, efficiency increase: ≈ 3%). Moreover, it is obvious that next to the heat transfer coefficients, process parameters do have a strong influence on the required heat exchange area. Therefore, it cannot be stated that lower heat transfer coefficients lead to higher heat transfer areas than comparing a certain application. It is always necessary to calculate the respective areas under consideration of the (sometimes even just slightly) different operating conditions. However, it can be concluded that siloxanes, as working fluids for high temperature applications, exhibit reduced heat transfer degradation compared to low temperature fluids like for example R245fa/R365mfc. As a result, heat transfer areas are not as additionally increased as in case of geothermal applications and, therefore, the use of mixtures can be beneficial, even though efficiency increases are comparably small.

In order to definitely decide whether the use of mixtures is economically feasible in a certain application, it is certainly necessary to analyse the heat transfer areas of internal heat exchanger and condenser as well. In addition, the performance of turbine and pump has to be investigated, too. Finally, specific problems that can occur by using mixtures, as for example compositions shifts during the process, have to be analysed in detail. Therefore, in future work, analogous investigations (measurements of heat transfer coefficients, identifying of appropriate correlations as well as estimation of required heat transfer areas) will be performed with respect to the condenser. Then, a holistic evaluation of using mixtures in ORC-applications is possible.

ACKNOWLEDGMENTS

The work was partially funded by the Deutsche Forschungsgemeinschaft (DFG) with project No. BR 1713/12. The authors gratefully acknowledge this support.

AUTHOR CONTRIBUTIONS

All authors contributed to this work by collaboration. Theresa Weith is the main author of this manuscript. Florian Heberle assisted in the conceptual design of the study as well as in the experiments. Markus Preißinger particularly contributed to the heat exchanger model. The whole project was supervised by Dieter Brüggemann. All authors revised and approved the publication.

CONFLICTS OF INTEREST

The authors declare no conflict of interest.

REFERENCES

1. Heberle, F.; Preißinger, M.; Brüggemann, D. Zeotropic mixtures as working fluids in Organic Rankine Cycles for low-enthalpy geothermal resources. Renew. Energy **2012**, 37, 364–370.

2. Aghahosseini, S.; Dincer, I. Comparative performance analysis of low-temperature Organic Rankine Cycle (ORC) using pure and zeotropic working fluids. Appl. Therm. Eng. **2013**, 54, 35–42.

3. Liu, Q.; Duan, Y.; Yang, Z. Effect of condensation temperature glide on the performance of organic Rankine cycles with zeotropic mixture working fluids. Appl. Energy **2014**, 115, 394–404.

4. Preißinger, M.; Heberle, F.; Brüggemann, D. Advanced Organic Rankine Cycle for geothermal application. Int. J. Low Carbon Technol. **2013**, 8, i62–i68.

5. Chen, H.; Goswami, D.; Rahman, M.; Stefanakos, E. A supercritical Rankine cycle using zeotropic mixture working fluids for the conversion of low-grade heat into power. Energy **2011**, 36, 549–555.

6. Baik, Y.-J.; Kim, M.; Chang, K.-C.; Lee, Y.-S.; Yoon, H.-K. Power enhancement potential of a mixture transcritical cycle for a low-temperature geothermal power generation. Energy **2012**, 47, 70–76.

7. Yang, K.; Zhang, H.; Wang, Z.; Zhang, J.; Yang, F.; Wang, E.; Yao, B. Study of zeotropic mixtures of ORC (organic Rankine cycle) under engine various operating conditions. Energy **2013**, 58, 494–510.

8. Zhang, J.; Zhang, H.; Yang, K.; Yang, F.; Wang, Z.; Zhao, G.; Liu, H.; Wang, E.; Yao, B. Performance analysis of regenerative organic Rankine cycle (RORC) using the pure working fluid and the zeotropic mixture over the whole operating range of a diesel engine. Energy Convers. Manag. **2014**, 84, 282–294.

9. Angelino, G.; Colonna, P. Multicomponent Working Fluids for Organic Rankine Cycles (ORCs). Energy **1998**, 23, 449–463.

10. Angelino, G.; Colonna, P. Air cooled siloxane bottoming cycle for molten carbonate fuel cells. In Proceedings of the Fuel Cell Seminar, Portland, OR, USA, 30 October 2000; pp. 667–670.

11. Dong, B.; Xu, G.; Cai, Y.; Li, H. Analysis of zeotropic mixtures used in high-temperature Organic Rankine cycle. Energy Convers. Manag. **2014**, 84, 253–260.

12. Chys, M.; van den Broek, M.; Vanslambrouck, B.; de Paepe, M. Potential of zeotropic mixtures as working fluids in organic Rankine cycles. Energy **2012**, 44, 623–632.

13. Kedzierski, M.; Kim, J.; Didion, D. Causes of the apparent heat transfer degradation for refrigerant mixtures. In Proceedings of the ASME/AIChe/ANS National Heat Transfer Conference, San Diego, CA, USA, 9–12 August 1992; pp. 149–158.

14. Stephan, K. Heat transfer in boiling of mixtures. In Proceedings of the Seventh International Heat Transfer Conference, Munich, Germany, 6–10 September 1982; pp. 59–81.

15. Bivens, D.; Yokozeki, A. Heat transfer of refrigerant mixtures. In Proceedings of the International Refrigeration and Air Conditioning Conference, Purdue University, West Lafayette, IN, USA, 14–17 July 1992; pp. 141–148.

16. Jung, D.; Song, K.; Ahn, K.; Kim, J. Nucleate boiling heat transfer coefficients of mixtures containing HFC32, HFC125, and HFC134a. Int. J. Refrig. **2003**, 26, 764–771.

17. Li, M.; Dang, C.; Hihara, E. Flow boiling heat transfer of HFO1234yf and R32 refrigerant mixtures in a smooth horizontal tube: Part I. Experimental investigation. Int. J. Heat Mass Transf. **2012**, 55, 3437–3446.

18. Choi, T.; Kim, Y.; Kim, M.; Ro, S. Evaporation heat transfer of R-32, R-134a, R-32/134a, and R-32/125/134a inside a horizontal smooth tube. Int. J. Heat Mass Transf. **2000**, 43, 3651–3660.

19. Tibiriçá, C.; Ribatski, G. Flow boiling heat transfer of R134a and R245fa in a 2.3 mm tube. Int. J. Heat Mass Transf. **2010**, 53, 2459–2468.

20. Ong, C.; Thome, J. Flow boiling heat transfer of R134a, R236fa and R245fa in a horizontal 1.030 mm circular channel. Exp. Therm. Fluid Sci. **2009**, 33, 651–663.

21. Fernández, F.; Prieto, M.; Suárez, I. Thermodynamic analysis of high-temperature regenerative organic Rankine cycles using siloxanes as working fluids. Energy **2011**, 36, 5239–5249.

22. Lai, N.; Wendland, M.; Fischer, J. Working fluids for high-temperature organic Rankine cycles. Energy **2011**, 36, 199–211.

23. Heberle, F.; Preißinger, M.; Weith, T.; Brüggemann, D. Experimental Investigation of Heat Transfer Characteristics and Thermal Stability of Siloxanes. Available online: http://www.asme-orc2013.nl/uploads/File/PPT%20082.pdf (accessed on 21 June 2014).

24. The Dow Chemical Company. Dowtherm G, Product Information. Available online: http://msdssearch.dow.com/PublishedLiteratureDOWCOM/dh_0033/0901b803800337f0.pdf?filepath=heattrans/pdfs/noreg/176-01466.pdf&fromPage=GetDoc (accessed on 19 May 2014).

25. Aspen Technology, Incorporation. Aspen One V 7, Process Optimization for Engineering, Manufacturing, and Supply Chain, Aspen Plus V 7.3; Aspen Technology: Burlington, ON, Canada, 2011.

26. Aspen Technology, Incorporation. Aspen Physical Property System. Physical Property Methods. Documentation to Aspen Plus V 7.3; Aspen Technology: Burlington, ON, Canada, 2011.

27. Lemmon, E.W.; Huber, M.L.; McLinden, M.O. NIST Standard Reference Database 23: Reference Fluid Thermodynamic and Transport Properties-REFPROP, Version 9.1; National Institute of Standards and Technology, Standard Reference Data Program: Gaithersburg, MD, USA, 2013.

28. Weith, T.; Heberle, F.; Brüggemann, D. Experimental Investigation of Flow Boiling Characteristics of Siloxanes and Siloxane Mixtures in a Horizontal Tube. In Proceedings of the International Symposium on Convective Heat and Mass Transfer, CONV-14, Kusadasi, Turkey, 8–13 June 2014.

29. Kandlikar, S. Boiling heat transfer with binary mixtures: Part II—Flow boiling in plain tubes. J. Heat Transf. **1998**, 120, 388–394.

30. Podhorsky, M.; Krips, H. Wärmetauscher—Aktuelle Probleme der Konstruktion und Berechnung, 2nd ed.; Vulkan-Verlag: Essen, Germany, 1999. (In German)

31. Wolf, P.; Kirchner, G.O. Konstruktive Hinweise für den Bau von Wärmeübertragern. In VDI-Wärmeatlas, 10th ed.; Verein Deutscher Ingenieure: Berlin, Germany, 2013. (In German)

32. Roetzel, W.; Spang, B.C. Berechnung von Wärmeübertragern. In VDI-Wärmeatlas, 10th ed.; Verein Deutscher Ingenieure: Berlin, Germany, 2013. (In German)

33. Gnielinski, V. Ein neues Berechnungsverfahren für die Wärmeübertragung im Übergangsbereich zwischen laminarer und turbulenter Rohrströmung. Forsch. Geb. Ing. **1995**, 61, 240–248. (In German).

34. MathWorks, Inc. MATLAB R2010b; MathWorks, Inc.: Natick, MA, USA, 2010.

35. Heberle, F. Untersuchungen zum Einsatz von zeotropen Fluidgemischen im Organic Rankine Cycle für die geothermische Stromerzeugung. Ph.D. Thesis, Universität Bayreuth, Bayreuth, Germany, 2013. (In German).

36. Heberle, F.; Brüggemann, D. Pool boiling heat transfer coefficients of R245fa, R365mfc and their mixtures. In Proceeding of the 8th World Conference on Experimental Heat Transfer, Fluid Mechanics, and Thermodynamics, Lisbon, Portugal, 16–20 June 2013.

4
CHAPTER

HEAT TRANSFER ENHANCEMENT VIA COMBINED WALL AND TRIANGULAR ROOTED-FIN SYSTEM

Abdul Rahim A. Khaled*, Abdullatif A. Gari

Mechanical Engineering Department, King Abdulaziz University, Jeddah, KSA

ABSTRACT

This work analyzes heat transfer through a wall containing triangular fins partially embedded in its volume. The coupled heat diffusion equations governing each constituent are solved numerically using an iterative finite volume method. A well bracketed effectiveness of the combined system that suits wide range of applications is analytically derived. Good agreement between the numerical and the analytical results is attained. It is found that the fin-root can act simultaneously as a heat sink and heat source for the wall. The heat transfer rate through the combined system is clearly seen to be maximized at a specific fin-root length. The maximum reported heat transfer rate through the triangular rooted-finned wall is found to be at most 90% above that for the rootless fin case at wall Biot number of 1.54. This percentage is noticed to decrease as the wall Biot number decreases. At that Biot number, the maximum heat transfer rate through the combined system reaches 150% above that through the plain wall. As a result of this work, it is recommended to utilize the triangular rooted-fin as a heat transfer enhancer for high mechanical strength structures exposed to highly convective fluid streams.

Keywords: Conduction; Convection; Fins; Augmentation; Roots; Thermal System

1. INTRODUCTION

Heat transfer through finned surfaces, like finned tubes, was widely studied in the literature [1] -[4] . Special interest to the present work is the wall that contains fins partially embedded in its volume. The rooted-finned wall systems can be proposed to be used in engineering applications including boiling, superheating and cooling of nuclear reactors. In these applications, the heating/cooling device must possess both high mechanical strength and high heat transfer capability. In order to satisfy both requirements, the main construction system of the device which can be the tube must be made from a high mechanical strength material like a stainless steel. As such, the heating/cooling device can support the high pressures of the fluid streams. Meanwhile, fins made of a large thermal conductivity material like copper can be partially embedded in the main construction system. The fin materials are generally not mechanically strong. As such, rooting the fin is necessary in order to enhance heat transfer through the less conductive main construction system. This is because these systems usually possess smaller thermal conductivities.

To the authors' best knowledge, the thermal performance of the rooted-finned wall systems was first studied by Khaled [5] [6] and Khaled and Al-qadi [7] . They have shown that the wall comprising rectangular rooted-fins may better augment heat transfer than the regular finned surface. This depends on the fin-root length, relative system dimensions, thermal conditions, and wall-to-fins thermal conductivities ratio. The rooted-finned wall system is expected to have better thermal performance if the fin-root has larger volume with larger interface area and when the external-fin portion has smaller volume with larger surface area. This is because the former features produce minimum wall conduction resistance while the latter ones increase the fin efficiency and decrease the thermal-convection resistance. The typical fin geometry that satisfies these features is the triangular fin with its base embedded inside the wall. The previous proposal received no attention in the heat transfer literature, and it is therefore the main motivation behind the present work.

In the next section, the two-dimensional heat diffusion equations governing the wall and the triangular rooted-fin materials are appropriately non-dimensionalized. Then, the corresponding boundary conditions and different suitable performance indicators are recognized. Next, an approximate simplified engineering model is casted for the first time in this work and a bracketed equation for the triangular rooted-finned wall effectiveness is identified. After that, the numerical and the approximate analytical computations of the performance indicators are compared. Eventually, an extensive parametric study is performed in order to fully characterize the thermal performance of the proposed system. Finally, the main, concluding remarks, are emphasized.

2. PROBLEM FORMULATION

2.1. General Model

Consider a two dimensional slab wall having a uniform thickness W and height H. The x-axis is taken along the wall thickness starting from the left surface while the y-axis is aligned along its height starting from the wall center-line as shown in Figure 1(a). Next, consider an isosceles triangular fin of length W and base thickness t. The fin-base is considered to be embedded inside the wall volume as shown in Figure 1(b). The distance between the wall right surface $(x=W)$ and the fin-base is L_o as seen from Figure 1(a). The x_f-axis of the fin is directed along the fin length starting from $x=x_o=W-L_o$ while the y_f-axis of the fin is aligned along the fin thickness starting from the fin center line at $y_f=y=0$ as shown in Figure 1(a). The thermal conductivity of the wall is k while that of the fin is k_f where $k_f>k$. Assume that heat transfer inside both materials is governed by the Fourier's law of conduction. The two-dimensional heat diffusion equations for both volumes are:

$$(a_H)^2\frac{\partial^2\theta}{\partial X^2}+\frac{\partial^2\theta}{\partial Y^2}=0 \tag{1}$$

$$(a_t)^2\frac{\partial^2\theta_f}{\partial X_f^2}+\frac{\partial^2\theta_f}{\partial Y_f^2}=0 \tag{2}$$

where

$$\theta=\frac{T(x,y)-T_{\infty}}{T_L-T_{\infty}};\quad \theta_f=\frac{T_f\left(x_f,y_f\right)-T_{\infty}}{T_L-T_{\infty}};$$
$$X=\frac{x}{W};\quad X_f=\frac{x_f}{W};\quad Y=\frac{y}{H};$$
$$Y_f=\frac{y_f}{t};\quad a_H=\frac{H}{W};\quad a_t=\frac{t}{W} \tag{3}$$

The quantities T, T_f, T_L and T¥ are the wall temperature field, fin temperature field, left surface temperature and the fluid free stream temperature, respectively.

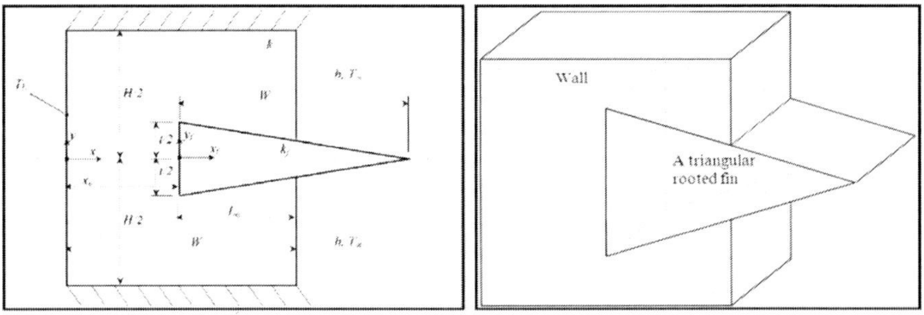

(a) (b)

Figure 1. (a) A 2D schematic diagram for a wall segment containing a triangular rooted-fin and the corresponding coordinates systems; (b) A 3D sketch for a wall containing a triangular rooted-fin.

The boundary conditions of Equations (1) and (2) are taken as:

$$\theta(X=0,Y)=1;\quad \left.\frac{\partial\theta}{\partial Y}\right|_{X,Y=-\frac{1}{2}}=\left.\frac{\partial\theta}{\partial Y}\right|_{X,Y=\frac{1}{2}}=0 \tag{4a,b}$$

$$X=1,\quad Y<-\frac{1}{2}X_o\left(\frac{a_t}{a_H}\right)\Rightarrow\frac{\partial\theta}{\partial X}+Bi\theta=0 \tag{4c}$$

$$X=1,\quad Y>\frac{1}{2}X_o\left(\frac{a_t}{a_H}\right)\Rightarrow\frac{\partial\theta}{\partial X}+Bi\theta=0 \tag{4d}$$

$$\left.\begin{array}{l} X = X_o, \quad -\frac{1}{2}\left(\frac{a_t}{a_H}\right) \le Y \le \frac{1}{2}\left(\frac{a_t}{a_H}\right) \\ X_f = 0, \quad -\frac{1}{2} \le Y_f \le \frac{1}{2} \end{array}\right] \Rightarrow \theta = \theta_f; \quad -k\frac{\partial \theta}{\partial X} = -k_f \frac{\partial \theta_f}{\partial X_f}$$

(5a,b)

$$X_f = 1, \quad Y_f = 0: \quad \frac{\partial \theta_f}{\partial X_f} = 0$$

(5c)

$$X_f > 1 - X_o, \quad Y_f = \mp\frac{1}{2}\left(1 - X_f\right): \mp\frac{\partial \theta_f}{\partial Y_f} + \frac{a_t^2}{2}\frac{\partial \theta_f}{\partial X_f} + \left(\frac{k}{k_f}\right)\left(a_t\sqrt{a_t^2/4+1}\right)Bi\theta_f = 0$$

(5d,e)

$$\left.\begin{array}{l} X_o \le X \le 1.0, \quad Y = \pm\frac{1}{2}\left(\frac{a_t}{a_H}\right)\left(1 + X_o - X\right) \\ 0 \le X_f \le 1 - X_o, \quad Y_f = \pm\frac{1}{2}\left(1 - X_f\right) \end{array}\right] \Rightarrow \theta = \theta_f$$

(6a,b)

$$\left.\begin{array}{l} X_o \le X \le 1.0, \quad Y = \mp\frac{1}{2}\left(\frac{a_t}{a_H}\right)\left(1 + X_o - X\right) \\ 0 \le X_f \le 1 - X_o, \quad Y_f = \mp\frac{1}{2}\left(1 - X_f\right) \end{array}\right] \Rightarrow \left(\frac{k_f}{k}\right)\left\{\left(\frac{a_t^2}{2}\right)\left(\frac{\partial \theta}{\partial X_f}\right) \mp \frac{\partial \theta}{\partial Y_f}\right\} = \left(\frac{a_t^2}{2}\right)\left(\frac{\partial \theta}{\partial X}\right) \mp \left(\frac{t}{H}\right)\left(\frac{\partial \theta}{\partial Y}\right)$$

(6c,d)

where X_o and Bi (wall Biot number) are equal to the following:

$$X_o = \frac{x_o}{W}; \quad Bi = \frac{hW}{k}$$

(7,8)

The quantity h is the convection heat transfer coefficient of the fluid stream. When the flow is made parallel to the depth of Figure 1(a), one can expect that all convective surfaces will have the same convective heat transfer coefficient. The total heat transfer rate through the left surface per unit depth represents the total heat transfer rate through the triangular rooted-finned wall. It can be evaluated from the following expression:

$$q'_L = -ka_H\left(T_L - T_\infty\right)\int_{-0.5}^{0.5} \left.\frac{\partial \theta}{\partial X}\right|_{X=0,Y} \mathrm{d}Y$$

(9)

The triangular rooted-finned wall effectiveness [7] is defined as the heat transfer rate through the combined system divided by that for the plain wall case. Mathematically, it is given by:

$$\varepsilon_W \equiv \frac{q'_L}{(q'_L)_{\text{no-fin}}} = \frac{q'_L}{hH\left[(T_R)_{\text{no-fin}} - T_\infty\right]} \tag{10}$$

When there is no fin, the wall right surface temperature, $(T_R)_{\text{no-fin}}$, can be obtained from the following expression using the one-dimensional heat transfer model [8] :

$$\frac{(T_R)_{\text{no-fin}} - T_\infty}{T_L - T_\infty} = \frac{1}{1+Bi} \tag{11}$$

As such, ew can be computed from the following expression:

$$\varepsilon_W = \left(1+\frac{1}{Bi}\right)\left\{-\int_{-0.5}^{0.5} \left.\frac{\partial \theta}{\partial X}\right|_{X=0,Y} \mathrm{d}Y\right\} \tag{12}$$

Define the rooted-finned wall second performance indicator γ as the ratio of the total heat transfer rate to that for the rootless fin case $(X_o = 1)$. It can be expressed as:

$$\gamma \equiv \frac{q'_L}{(q'_L)_{X_o=W}} = \frac{\varepsilon_W}{(\varepsilon_W)_{X_o=1}} \tag{13}$$

2.2. Approximate Solution

In spite of the complexity of the present system, ε_W may be bracketed by considering the present system as a piecewise one-dimensional combined system of Figure 2. The analyzed system is modeled as an arrangement of nine series and parallel different one-dimensional layers and the fluid stream medium. The thermal conductivity of layers A, B1, B2, D1 and D2 is k while that of layers C1, C2, E and the external-fin portion is k_f. The dimensions of each layer are clearly seen on Figure 2. The layers C1 and C2 are connected in series with layers D1 and D2, respectively. The layers B1,

B2, E, equivalent series C1-D1 layer and equivalent series C2-D2 layer are considered parallel layers. The latter equivalent parallel layers are connected in series with layer A to form the combined wall and fin-root system. This combined system is bounded by $0 \le x \le W$ and $-H/2 \le y \le H/2$. The right surface of the latter system located at $x = W$ is subjected to convection heat transfer over an area of $H - t(x_o/W)$ per unit depth and conduction heat transfer through the external-fin base over an area of $t(x_o/W)$ per unit depth.

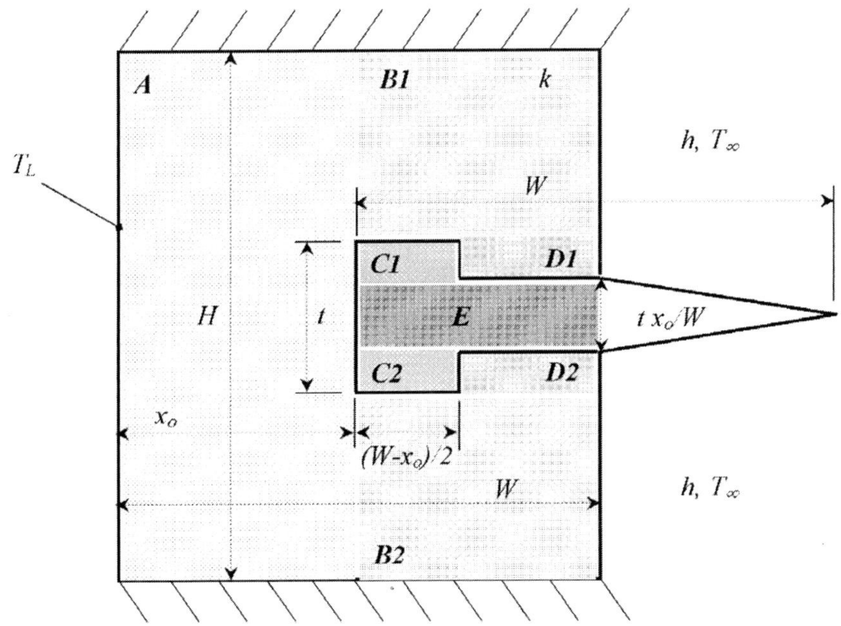

Figure 2. One dimensional piecewise model for a wall segment containing a triangular rooted-fin.

Recall that the thermal resistance [8] of a given layer of thickness W_L, height H_L, unit depth length, and thermal conductivity k_L is equal to: $R_{t,L} = W_L/(k_L H_L)$. The thermal resistance of the external-fin portion per unit depth is given by: $R_{t,f}=1/\left(h\eta_f 2x_o\sqrt{1+a_t^2}\right)$ where η_f is the efficiency of the external-fin portion.

Meanwhile, the convective thermal resistance of the right isothermal surface of the wall per unit depth is equal to: $R_{t,c}=1/\left(h[H-tX_o]\right)$. Also, recall that the equivalent thermal resistance $(R_{t,e})$ of n-series thermal resistances $(R_{t,i}, i=1,2,\cdots,n)$ is computed from the following formula: $R_{t,e}=\sum_{i=1}^{n}R_{t,i}$. On the other hand, $R_{t,e}$ of n-parallel thermal resistances $(R_{t,i}, i=1,2,\cdots,n)$ is obtained from the following formula: $1/R_{t,e}=\sum_{i=1}^{n}\left(1/R_{t,i}\right)$.

Finally, recall that the total heat transfer rate through the system shown in Figure 2 can be calculated from the following expression: $q_L'=(T_L-T_\infty)/R_{\text{total}}$ where R_{total} is the total equivalent thermal resistance between the left surface $(x=0)$ and the fluid free-stream on the right side of the system.

Using Equation (10) and the previous recalled formulations, the triangular rooted-finned wall effectiveness ε_W can be bracketed into the following form:

$$\varepsilon_W=\frac{1+Bi}{S+X_oBi+\dfrac{(a_H/a_t)(k/k_f+1)(1-X_o)Bi}{(1-k/k_f)+X_o(k_f/k-1)+(a_H/a_t)(k/k_f+1)}} \qquad (14)$$

where S is given by:

$$S=\frac{a_H}{a_H-a_tX_o+2\eta_fX_o\sqrt{1+a_t^2/4}} \qquad (15)$$

Using results available in any of heat transfer textbooks [8] , η_f for the external-fin portion can be obtained from the following correlation:

$$\eta_f=-0.093478Z^5+0.44565Z^4-0.64053Z^3+0.21691Z^2-0.22228Z+1 \qquad (16)$$

where Z based on the external-fin portion geometry is given by:

$$Z = \sqrt{2\left(\frac{k}{k_f}\right)\left(\frac{X_o}{a_t}\right)Bi} \tag{17}$$

3. NUMERICAL METHODOLOGY AND VALIDATION

3.1. Numerical Methodology

The two-dimensional heat diffusion equations given by Equations (1) and (2) are coupled via their boundary conditions represented by Equations (4), (5) and (6). These systems of equations have been solved using an iterative finite volume method [9] . A third order scheme is used to discretize these equations [10] in order to reduce the numerical diffusion of errors. The selected convergence criterion requires that the sum of the energy equations residues between the current and the previous solutions be less than 10^{-12}. Furthermore, the solution convergence has been ensured by checking the value of heat transfer rate at the left surface $(x=0)$ and the sum of the heat transfer rate at the wall right surface $(x=W)$ and heat transfer rate from the fin surfaces exposed to the fluid stream. For all runs, the maximum sum of these values was found to be less than 10^{-5}. As such, the conservation of energy was satisfied, $\dot{E}_{in} = \dot{E}_{out}$. Furthermore, a fine mesh of 16,700 elements was selected. When a numerical result of one case based on that mesh was compared with the result based on a mesh of 4175 elements, the percentage difference in ε_W was found to be less than 0.02%. As such, solutions independent on the grid size was attained. It should be mentioned that for the situation where the interface between the fin-root and the wall is not a perfect contact surface, other methods, such as lattice Boltzmann method may be used [11] .

3.2. Validation of the Numerical Solution

Comparisons between the numerical results (2D solution) and analytical results (1D solution, Equation (14)) are shown in Figures 3 and 4. Excellent agreement are seen when $X_o \geq 0.3$ for all Bi numbers and when

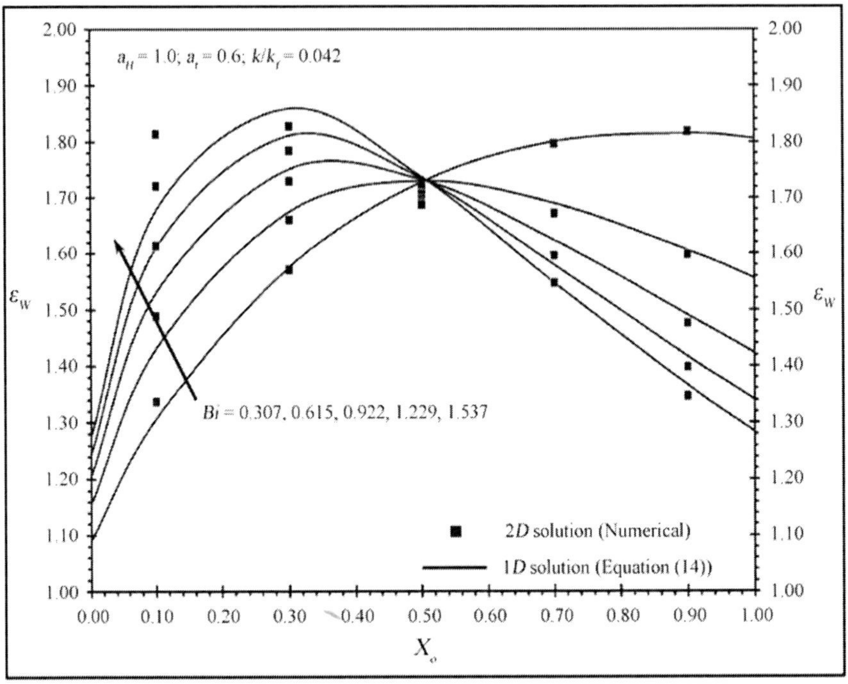

Figure 3. Effects of X_o and Bi on ε_W.

$X_o \leq 0.1$ if $Bi \leq 0.5$. This led to more confidence on the obtained results. Moreover, the following validity condition of the present 1D solution is recommended:

$$Bi(1 - X_o) = \frac{hL_o}{k} < 0.5 \tag{18}$$

The condition given by Equation (18) requires that the variation in the temperature of the wall across the root length and the variation in the temperature of the fin-root are negligible. This indicates negligible transverse conduction across the wall-root interface as compared to the axial conduction through the fin root-base. As such, the analytical 1D model given by Equation (14) is valid.

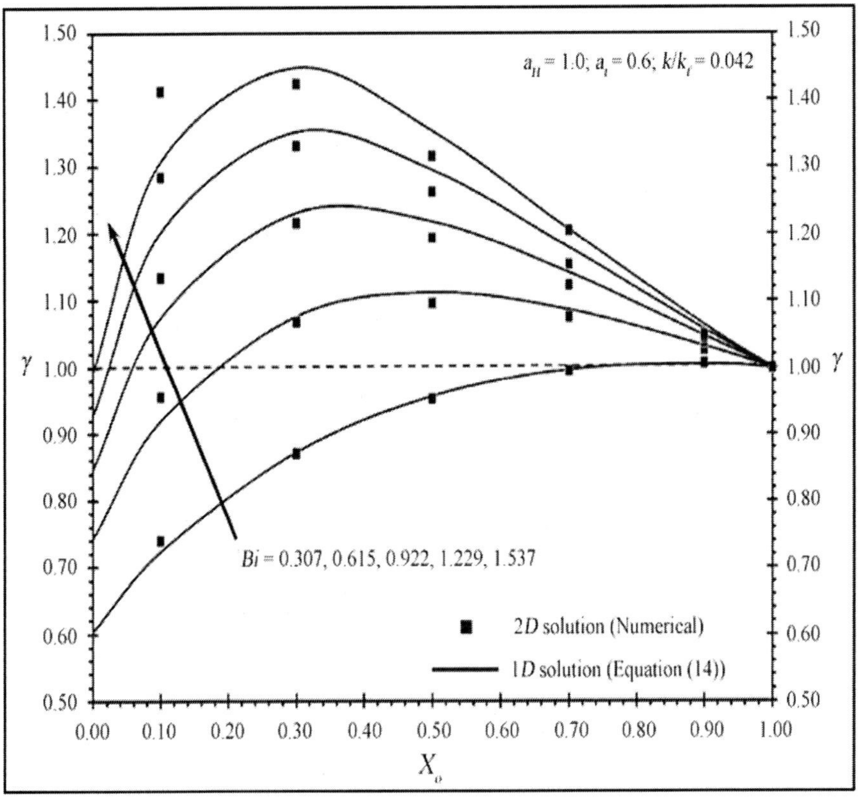

Figure 4. Effects of X_o and Bi on γ.

4. DISCUSSION OF THE RESULTS

4.1. Discussion of Temperature Contour Plots

Figure 5 shows temperature contours inside the different constituents of the present system at two different Biot numbers with various root-base positions. The temperature contours inside the fin-root and the wall show existence of transverse heat fluxes originated from the wall material around the root-base and directed toward the fin-root. The magnitude of these fluxes can be noticed to decrease as X_o increases until it is eliminated when $X_o = 1$. Meanwhile, it can be depicted from these plots that there can be heat fluxes originated from the fin-root close to the fin-base and directed toward the wall right surface. This is because the wall two sharp edges surrounding the fin-base have negligible conduction resistances. The previous phenomenon is noticed to disappear when $X_o \geq 0.9$. The two dimensional induced heat fluxes are expected to be negligible when $Bi(1 - X_o) \leq 0.5$ as discussed earlier.

4.2. Discussion of System Effectiveness and Second Performance Indicator

As the root length increases, the wall conduction resistance decreases while the total convection resistance increases. This phenomenon is capable of minimizing the total thermal resistance of the combined system at a

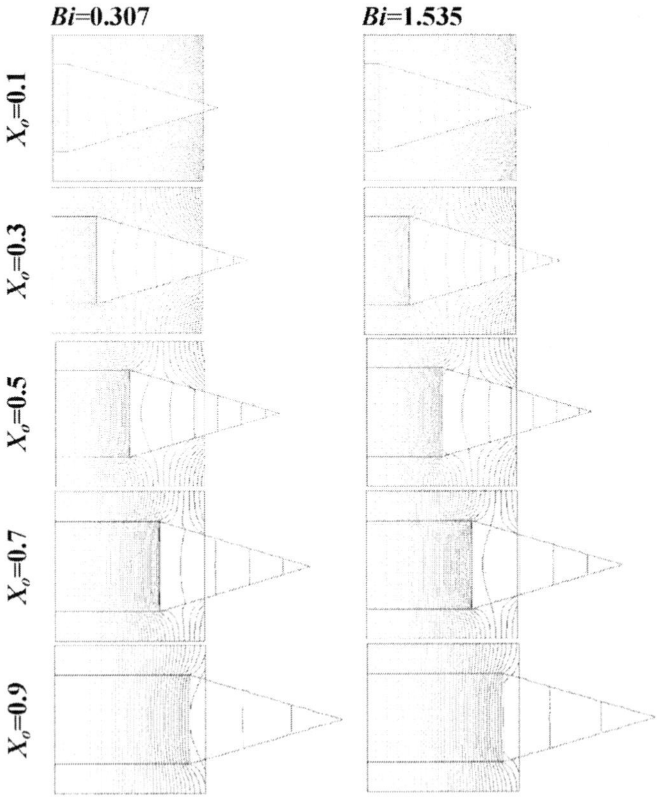

Figure 5. Effects of X_o and Bi on temperature contours for a_H = 1.0, a_t = 0.6, and k/k_f = 0.042.

specific root length. As such, the total heat transfer rate can be maximized at this length. Figure 3 demonstrates the previous phenomenon. The maximum system effectiveness seen in Figure 3 is noticed to vary with Bi as this quantity clearly alters in the total thermal resistance of the combined system. It is seen from Figure 3 that ε_W becomes independent on Bi at a specific dimensionless fin-root length which is equal to $(1-X_o)=0.48$. This means that both q'_L and $(q'_L)_{\text{no-fin}}$ increase by the same extent with an increase in Bi at that specific fin-root length.

Figure 4 shows that the maximum value of γ is $\gamma_{\max}=1.449$ when $Bi=1.535$. This means that the maximum heat transfer rate when $Bi=1.535$ is 44.9% above that for the rootless fin case. This maximum value occurs at a dimensionless fin-root length of $(1-X_o)=0.7$. $\gamma_{\max}$ is seen to increase as Bi increases as shown in Figure 4. Finally, Figure 6 shows the variation of the maximum combined system effectiveness $\varepsilon_{W,\max}$ and maximum second performance indicator $\gamma_{\max}$ at $Bi=1.535$ as functions of both a_t and k/k_f values. This figure shows that $\gamma_{\max}$ becomes almost independent on a_t when $k_f\to\infty$. According to this figure, the maximum heat transfer rate through the triangular rooted-finned wall increases until it approches asymptotically 90% above that for the rootless fin case at $Bi=1.535$.

5. CONCLUSIONS

The problem of computing heat transfer rate through a system composed of a wall and triangular rooted-fin is theoretically investigated. Appropriate forms of the coupled heat diffusion equations for the wall and the triangular rooted-fin systems were solved using a finite volume method. Approximate closed form solution for the present problem based on a wide range of engineering applications was derived. Good agreement was obtained between the numerical and the closed form solutions when the wall Biot number based on the fin-root length is smaller than the one-half. The following remarks were concluded: 1) there is a critical fin-root length that maximizes the total heat transfer rate; 2) there is a critical fin-root length that makes the total system effectiveness independent on Biot number; 3) the maximum heat transfer rate through the combined system relative to that for the rootless fin case increases as the wall Biot number increases; 4) the maximum heat transfer rate through the triangular rooted-finned wall can be at most 90% above that for the rootless fin case at $Bi = 1.535$; and 5) the maximum heat transfer rate for the combined system can reach 150% above that through the plain wall at $Bi = 1.535$. Finally, it is recommended to consider triangular rooted-fins as heat transfer enhancers for highly mechanical strength structures exposed to highly convective fluid streams.

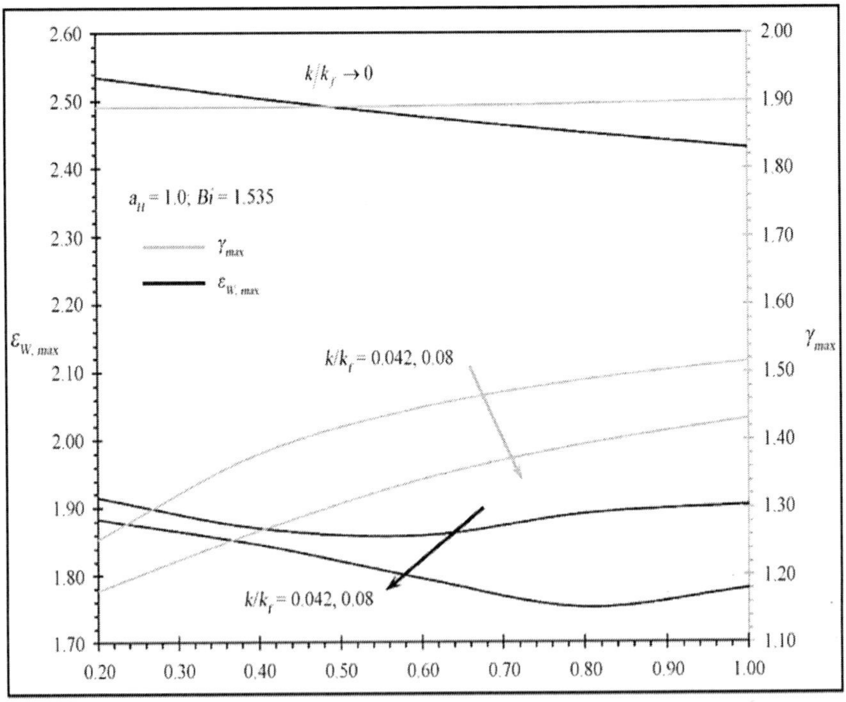

Figure 6. Effects of a_t and k/k_f on $\varepsilon_{w,max}$ and γ_{max}.

ACKNOWLEDGEMENTS

The authors acknowledge the full support of this work by King Abdulaziz City for Science and Technology (KACST) under project no. 8-ENE192-3.

REFERENCES

1. Bergles, A.E. (1998) Handbook of Heat Transfer. 3rd Edition, McGraw-Hill, New York, 11.1-11.76.

2. Kraus, A.D., Aziz, A. and Welty, J.R. (2001) Extended Surface Heat Transfer. John Wiley & Sons, Inc., New York.

3. Goldstein, R.J., Ibele, W.E., Patankar, S.V., Simon, T.W., Kuehn, T.H., Strykowski, P.J., Tamma, K.K., Heberlein, J.V.R., Davidson, J.H., Bischof, J., Kulacki, F.A., Kortshagen, U., Garrick, S., Srinivasan, V., Ghosh, K. and Mittal, R. (2010) Heat transfer—A Review of 2004 Literature. International Journal of Heat and Mass Transfer, 53, 4343-4396.

4. Goldstein, R.J., Ibele, W.E., Patankar, S.V., Simon, T.W., Kuehn, T.H., Strykowski, P.J., Tamma, K.K., Heberlein, J.V.R., Davidson, J.H., Bischof, J., Kulacki, F.A., Kortshagen, U., Garrick, S., Srinivasan, V., Ghosh, K. and Mittal, R. (2010) Heat Transfer—A Review of 2005 Literature. International Journal of Heat and Mass Transfer, 53, 4397-4447.

5. Khaled, A.-R.A. (2008) Heat Transfer Analysis through Solar and Rooted Fins. Journal of Heat Transfer—Transactions of the ASME, 130, 74503.

6. Khaled, A.-R.A. (2011) Heat Transfer Enhancements through Rooted Rectangular-and-Triangular Fins. Journal of Enhanced Heat Transfer, 18, 127-136.

7. Khaled, A.-R.A. and Al-qadi, I. (2012) Augmentation of Heat Transfer in Wall-Rooted-Fins Systems. International Journal of Numerical Methods for Heat and Fluid Flow, 22, 194-214.

8. Incropera, F.P., Dewitt, D., Bergman, T.L. and Lavine, A.S. (2011) Fundamentals of Heat and Mass Transfer. 7th Edition, John Wiley & Sons, Inc., New York.

9. LeVeque, R. (2002) Finite Volume Methods for Hyperbolic Problems. Cambridge University Press, Cambridge.

10. van Leer, B. (1979) Towards the Ultimate Conservative Difference Scheme. V-A Second Order Sequel to Godunov's Method. Journal of Computational Physics, 32, 101-136.

11. Xiong, Q., Li, B., Zhoua, G., Fanga, X., Xua, J., Wanga, J., Hea, X., Wanga, X., Wanga, L. and Gea, W. (2012) Large-Scale DNS of Gas-Solid Flows on Mole-8.5. Chemical Engineering Science, 71, 422-430.

5
CHAPTER

CONDENSATION HEAT TRANSFER IN HORIZONTAL NON-CIRCULAR MICROCHANNELS

Hicham El Mghari[1,2], Mohamed Asbik[1], Hasna Louahlia-Gualous[2]

[1]LP2MS, URAC08, Université Moulay Ismaïl, Faculté des Sciences, Meknès, Maroc

[2]LUSAC, Université de Caen Basse Normandie, Saint Lô, France

ABSTRACT

This investigation contributes to a better understanding of condensation heat transfer in horizontal non-circular microchannels. For this purpose, the conservation equations of mass, momentum and energy have been numerically solved in both phases (liquid and vapor), and all the more, so the film thickness analytical expression has been established. Numerical results relative to variations of the meniscus curvature radius, the condensate film thickness, the condensation length and heat transfer coefficients, are analyzed in terms of the influencing physical and geometrical quantities. The effect of the microchannel shapes on the average Nusselt number is highlighted by studying condensation of steam insquare, rectangular and equilateral triangular microchannels with the same hydraulic diameter of 250 μm.

Keywords: Condensation; Microchannel; Numerical simulation; Capillary regime; Heat transfer

1. INTRODUCTION

Understanding the heat transfer behavior of condensation flow in microchannels is important for a broad variety of engineering applications. Although there have been a number of investigations on boiling flow in microchannels, there are relatively little experimental data and theoretical analyses relative to condensation processes available in the literature, especially, for condensation inside a noncircular microchannels. In the chapter 6 of the reference [1], an overview of minichannels and microchannels condensation has been exposed.

Most of the physical and mathematical models that focused on annular condensation heat transfer in circular channel were developed in the previous works. Begg et al. [2] studied annular film condensation in a small circular tube to predict the shape of the liquid-vapor interface along a miniature tube leading to the complete condensation phenomena in small diameter tubes. LouahliaGualous and Asbik [3] conducted a numerical model predicting heat transfer for condensation of pure refrigerant and binary mixture in a mini-tube. Miscevic et al. [4] developed a stationary condensation capillary flow model based on the separate flow approach by taking into account the coupling between a cylindrical interface and a hemispherical interface. Recently, Ribeiro et al. [5] experimentally investigated the thermal-hydraulic performance of microchannel condensers using three different copper metal foams structures with distinct pore densities and porosities (0.893 and 0.947) as enhanced surfaces on the air-side. Their results are compared with the conventional condenser surface. El Achkar et al. [6] investigated the experimental heat transfer in the isolated bubbles zone of a transparent circular cross-section micro condenser. The evolution of vapor quality was experimentally determined by using the image processing. The energy balance was then used to calculate the temperature of the liquid in the isolated bubbles zone, showing that liquid and vapor were not in thermal equilibrium. The sensible heat transfers and latent heat transfers were then compared. However, a fundamental understanding of local mechanisms of heat and mass transfer cannot be accomplished using multichannels because the mass flow rates for each microchannel and condensate flow regime are unknown. Also, condensation heat transfer and pressure drop depend on the corresponding structure of the two phase flow (mist flow, annular flow, bubbly flow, or slug/plug flow) as shown by Odaymet et al. [7]. Three main condensation flows were identified in a small circular tube: the annular flow, the intermittent or elongated bubbles flow, and the spherical bubbles flow by Louahlia-Gualous and Mecheri [8]. Annular flow is especially found to be one of the dominant condensation flows in microchannelsas shown by Odaymet and LouahliaGualous [9] and Quan et al. [10].

On the other hand, various theoretical models have been proposed to predict the local heat transfer related to the condensation annular flow in non-circular channel where the surface tension plays a predominant effect on the condensate flow, more specifically, in the channel corners. Indeed, Zhao and Liao [11] analyzed annular film condensation heat transfer inside vertical mini triangular channel using three zones: the thin liquid film flow on the sidewall, the condensate flow in the corners, and the vapor core flow in the center. Wu and Cheng [12] carried out a simultaneous visualization and measurement experiment to perform condensation flow patterns of steam flowing through an array of trapezoidal silicon microchannels with a hydraulic diameter of 82.8 μm. Wang et al. [13] and Wang & Rose [14] proposed a theoretical model for condensation annular flow in a horizontal square and equilateral triangular channel with hydraulic diameter ranging from 0.5 to 5 mm by taking into account the effects of gravity, surface tension, and interfacial shear stress. They obtained the local heat transfer coefficient for refrigerants R134a, R22, R152a, CO2, propane, ammonia, and R410a by assuming that the channel wall temperature is uniform. Furthermore, they proposed one correlation for condensation heat transfer in the square and triangular microchannels in which surface tension and viscosity are the predominant parameters controlling condensate film thickness. Additionally, in references ([15-18]) for which the contents are not detailed here, readers could find more information about the use of various non-circular microchannel shapes.

The main purpose of the present work is to determine the heat transfer coefficient during the steam condensation inside horizontal various non-circular microchannels (rectangle, square, or equilateral triangle). Indeed, the classical mathematical model of the annular condensation flow is retained in both phases (liquid and vapor). An appropriate numerical method is used to solve the differential equations system obtained from different conservation equations (mass, momentum and energy). To compute the heat transfer coefficient, a new and simple geometrical method is used to express the condensate film thickness.

2. PHYSICAL MODEL

The physical model investigated in this paper is illustrated on the Figure 1. It concerns, a horizontal non circular microchannel with a hydraulic diameter D, which is cooled with a uniform wall heat flux density q. At the microchannel inlet, the vapor (steam) mass flux, the pressure and temperature of the steam is given at the saturated state. The vapor condenses inside the microchan nel having wall temperature lower than vapor saturation temperature. The film thickness on the heat exchange surface varies along the axial direction with the vapor quality.

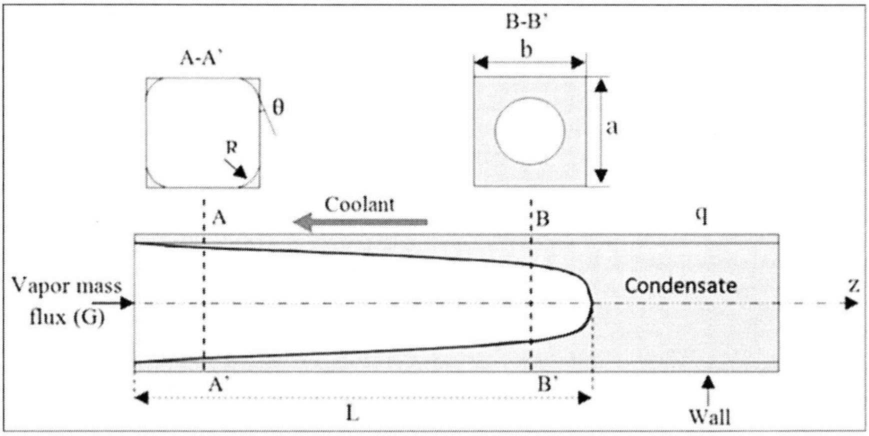

Figure 1. Schematic representation of the annular condensation in a rectangular microchannel.

When total condensation occurs, the end of the condensation zone has a hemispherical meniscus. Its length L is one of the unknown parameters in the physical model. The condensate film flows along the axial direction under effects of the pressure, surface tension, and shear stress.

3. MATHEMATICAL FORMULATION

3.1. Hypotheses

The mathematical formulation of the problem is based on the following principal assumptions:

- For liquid and vapor phases, thermophysical properties are assumed to be constant.
- In the microchannel, the flow is supposed to be steady-state, laminar, one-dimensional and axis-symmetrical.

- The free surface of the condensate film is smooth.
- Gravity forces are negligible compared to the effects of surface tension.
- The heat transfer from the cooling fluid to the condensate flow is assumed to be one-dimensional.
- In the condensate, the temperature profile is supposed linear.
- The saturation temperature of the vapor is assumed to remain constant along the microchannel.

3.2. Conservation Equations

The modeling approach developed here describes the liquid and vapor phases separately. The governing equations are used in the Cartesian coordinates as shown in the Figure 1.

3.2.1. Mass Conservation

The average parameters over a cross-section are used in liquid and vapor phases of the condensation flow respecting continuity conditions at the liquid-vapor interface. The equation of the mass conservation can be written for each local cross section as follows:

$$\frac{1}{A}\frac{\partial}{\partial z}\left(\rho_l U_l (1-\alpha) A\right) = -\Gamma \tag{1}$$

$$\frac{1}{A}\frac{\partial}{\partial z}\left(\rho_v U_v \alpha A\right) = \Gamma \tag{2}$$

where Γ represents the volumic rate of phase change. By convention, its sign is negative for condensation and positive for evaporation. So, we can write:

$$\Gamma = -\mathrm{d}\dot{m}_L / \mathrm{d}z = \mathrm{d}\dot{m}_v / \mathrm{d}z \tag{3}$$

A_ξ, A_ξ and U_ξ are respectively the microchannel cross section, the density and the axial velocity. The subscript ξ refers to the considered phase (ξ = v or L).

3.2.2. Momentum Conservation

For a fixed position z along the microchannel, the condensate film thickness in noncircular microchannels is much thicker into the microchannel corners than elsewhere, especially; at the internal considered circumference because of the surface tension effect. This is the reason that the axial flow in the film region between the corners is neglected.

From the forces balance illustrated on the Figure 2, the momentum conservation equations in the control volume of the length dz, are given by the equations (4) and (5). These relationships are essentially expressed in terms ofthe interfacial shear at the liquid-vapor interface, the shear wall friction at the liquid-wall contact surface, and the pressure forces on the liquid area. In both phases, all the physical properties are assumed to be constant, and the influences of the gravity and the buoyancy forces are neglected. So, in the liquid and vapor phases, the momentum conservation equations are:

- In the liquid phase:

$$-A_L \frac{\mathrm{d}P_L}{\mathrm{d}z} + \tau_{vL} \frac{\mathrm{d}S_{vL}}{\mathrm{d}z} - \tau_w \frac{\mathrm{d}S_{Lw}}{\mathrm{d}z} - U_L \frac{\mathrm{d}\dot{m}_L}{\mathrm{d}z} = \dot{m}_L \frac{\mathrm{d}U_L}{\mathrm{d}z} \tag{4}$$

- In the vapor phase:

$$-A_v \frac{\mathrm{d}P_v}{\mathrm{d}z} + \tau_{vL} \frac{\mathrm{d}S_{vL}}{\mathrm{d}z} - U_v \frac{\mathrm{d}\dot{m}_v}{\mathrm{d}z} = \dot{m}_v \frac{\mathrm{d}U_v}{\mathrm{d}z} \tag{5}$$

A_ξ, A_ξ and U_ξ are respectively the microchannel cross-section, the pressure and the axial velocities in the phase $\xi(v \text{ or } L)$. Furthermore, we indicate that $\mathrm{d}S_{vL}$ is the liquid–vapor interface surface along $\mathrm{d}z$, $\mathrm{d}S_{Lw}$ is the wet heat exchange surface along $\mathrm{d}z$, and τ_{vL} is the shear stress at liquid–vapor interface, τ_w is the shear stress at the microchannel heat exchange surface.

3.2.3. Energy Conservation

The local energy equation in the liquid phase can written as:

$$q\mathcal{P}z = \dot{m}_L(z)h_{fg} \tag{6}$$

The total energy equation defined in the length of the total condensation zone as:

$$q\mathcal{P}L = \dot{m}_{v,in} h_{fg} \tag{7}$$

where q is the heat flux density, $\dot{m}_L(z)$ is the local liquid mass flow rate $\mathcal{P}$ is the microchannel perimeter, and z is the microchannel abscissa.

3.2.4. Curvature Radius Expression

To express the curvature radius derivative, we need to use the Laplace–Young equation:

$$\frac{\mathrm{d}P_L}{\mathrm{d}z} = \frac{\mathrm{d}P_v}{\mathrm{d}z} + \frac{\sigma}{R^2}\frac{\mathrm{d}R}{\mathrm{d}z} \tag{8}$$

Combining equations (1)-(8), we get:

$$\frac{\mathrm{d}R}{\mathrm{d}z} = \frac{0.5 f_{vL}\rho_v \mathcal{P}_{vL} U_v^2 \left[1 + \frac{A_L}{A_v}\right] - 0.5 f_L \rho_L \mathcal{P}_{Lw} U_L^2 - \frac{2q\mathcal{P}A_L}{h_{fg}}\left[\frac{U_v}{A_v} + \frac{U_L}{A_L}\right]}{\sigma\frac{A_L}{R^2} + 2\frac{\rho_v U_v^2 A_L^2}{A_v R} - \frac{2\rho_L A_L}{R} U_L^2} \tag{9}$$

3.2.5. Dimensionless Equations

To establish the dimensionless equations, the following variables are used:

$$R^* = \frac{R}{D_h}, z^* = \frac{z}{D_h}, L^* = \frac{L}{D_h}, \bar{\rho} = \frac{\rho_v}{\rho_L}, P^* = \frac{\mathcal{P}}{D_h}, P^* = \frac{PD_h}{2\sigma}, U_V^* = \frac{A\rho_v U_v}{\dot{m}_{v,in}}, U_L^* = \frac{A\rho_L U_L}{\dot{m}_{v,in}} \tag{10}$$

Then the dimensionless form of the derivative curvature radius, the velocity gradients and the pressure gradients in both phases, are given by:

$$\frac{\mathrm{d}R^*}{\mathrm{d}z^*} = \frac{0.5 f_{vL} \mathcal{P}_{vL}^* U_v^{*2}\left[1 + \frac{A_L^*}{A_v^*}\right] - 0.5 f_L \rho^* \mathcal{P}_{Lw}^* U_L^{*2} - 2Bo\mathcal{P}^* A_L^*\left[\rho^* \frac{U_L^*}{A_L^*} + \frac{U_v^*}{A_v^*}\right]}{\frac{A_L^* \rho^*}{Re\,Ca} + \frac{2A_L^{*2} U_v^{*2}}{A_v^* R^*} - \frac{2\rho^* A_L^* U_L^{*2}}{R^*}} \tag{11}$$

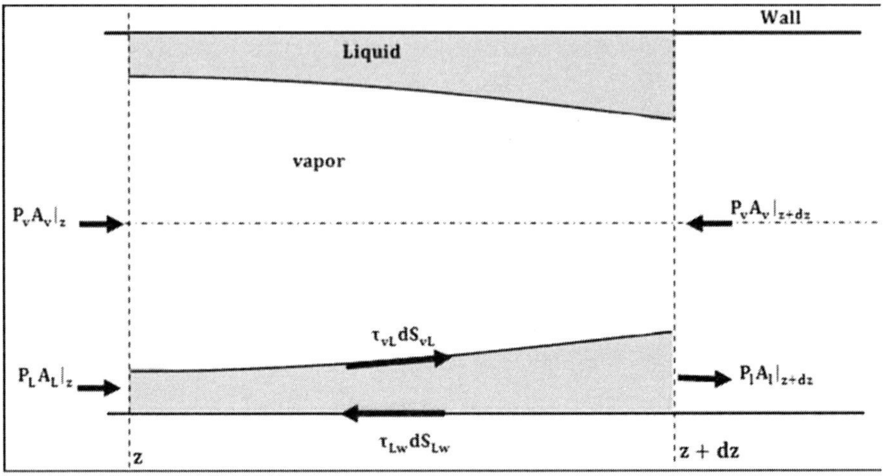

Figure 2. Forces balance for two-phase flow in a horizontal microchannel.

$$\frac{dU_L^*}{dz^*} = \frac{Bo\mathcal{P}^*}{A_L^*} - 2\frac{U_L^*}{R^*}\frac{dR^*}{dz^*} \tag{12}$$

$$\frac{dU_v^*}{dz^*} = -\frac{Bo\mathcal{P}^*}{A_v^*} - 2\frac{A_L^*}{A_v^*}\frac{U_v^*}{R^*}\frac{dR^*}{dz^*} \tag{13}$$

$$\frac{dP_L^*}{dz^*} = \frac{1}{4}\frac{f_{vL}Ca\ Re\mathcal{P}_{vL}^*U_v^{*2}}{\bar{\rho}A_L^*} - \frac{1}{4}\frac{f_L Ca\ Re\mathcal{P}_{Lw}^*U_L^{*2}}{A_L^*} - \frac{Bo\ Ca\ Re\mathcal{P}^*U_L^*}{A_L^*} + \frac{Ca\ ReU_L^{*2}}{R^*}\frac{dR^*}{dz^*} \tag{14}$$

$$\frac{dP_v^*}{dz^*} = -Ca\ ReU_v^*\frac{dU_v^*}{dz^*} + \frac{Ca\ Re}{A_v^*}\frac{dA_v^*}{dz^*} + \frac{Ca\ ReU_v^{*2}}{R^*}\frac{A_L^*}{A_v^*}\frac{dR^*}{dz^*} - \frac{1}{4}\frac{f_{vL}Ca\ Re\mathcal{P}_{vL}^*U_v^{*2}}{\bar{\rho}A_v^*} \tag{15}$$

$$Bo = q/Gh_{fg}\,,\quad Ca = \frac{\mu_l G}{\sigma \rho_L},$$

$$Re = \frac{\dot{m}_{v,in} D_h}{A\mu_L},\quad \dot{m}_{v,in} = GA$$

are respectively the boiling number, the capillary number, the vapor Reynolds number and the vapor mass flux.

Definitions of other parameters appearing in the above relations (11-15) are such that:

1) The liquid friction factor for laminar is given by:

$$f_L = CRe_L^{-1} \tag{16}$$

C is the Poiseuille number given in [19].

For turbulent flow, the liquid friction coefficient is determined from the Blasius equation [1]:

$$f_L = 0.0791\,\mathrm{Re}_L^{-0.25} \tag{17}$$

In this expression, the liquid Reynolds number Re_L is calculated assuming the liquid single phase:

$$Re_L = \frac{\rho_L U_L D_L}{\mu_L},\quad D_L = \frac{4A_L}{P_{Lw}} \tag{18}$$

2) The interfacial frictional coefficient taking into account the effect of the condensation process on the interfacial shear stress is defined by [20]:

$$f_{vL} = f_{v,0}\,\frac{\varnothing \mathrm{e}^{\varnothing}}{\mathrm{e}^{\varnothing} - 1} \tag{19a}$$

where $f_{v,0}$ is the friction factor for single phase vapor flow defined for laminar flow as :

$$f_{v,0} = CRe_v^{-1} \tag{19b}$$

For turbulent flow, $f_{v,0}$ is determined from:

$$f_{v,0} = 0.0791 Re_v^{-0.25} \tag{20a}$$

$$Re_v = \frac{\rho_v U_v D_v}{\mu_v}, \; D_v = \frac{4A_v}{\mathcal{P}_{vL}}, \tag{20b}$$

The factor $\varnothing$ is defined as the ratio of the local condensation mass flow rate to the vapor mass flow rate rebounding from the liquid-vapor interface [2], which is approximated by:

$$\varnothing = \frac{dq}{\mathcal{P}dz} \frac{2}{h_{fg} U_v \rho_v f_{v,0}} \tag{21}$$

3.2.6. Dimensionless Boundary Conditions

Equations (11) to (15) are solved using the following dimensionless boundary conditions:

1) At the microchannel inlet ($z^* = 0$):

- the flow mass flux G is imposed;
- the temperature and pressure at the saturated state are given;
- the non-dimensional vapor velocity is

$$U_v^*\left(z^* = 0\right) = 1;$$

• the non-dimensional liquid velocity is

$$U_L^*\left(z^* = 0\right) = 0 \; ;$$

• the non-dimensional curvature radius is

$$R^*\left(z^* = 0\right) = \frac{\sigma}{P_{v0} D_h}$$

2) At the position $z^* = L^*$ corresponding to the end of the condensation zone:

• the non-dimensional outlet liquid pressure is:

$$P_l\left(z^* = L^*\right) = \frac{P_0 D}{2\sigma};$$

• for the rectangular cross section, the dimensionless curvature radius is [22]:

$$R^*\left(z^* = L^*\right) = \varepsilon \sin(\beta) / \left(2\cos(\theta+\beta) D_h\right)$$

where $\varepsilon = \min(a,b)$;

• for the equilateral triangle cross section, the dimensionless curvature radius a is [18] :

$$R^*\left(z^* = L^*\right) = \mathcal{P} \sin(\beta) / \left(6\cos(\theta+\beta) D_h\right)$$

4. SOLUTION PROCEDURE

The above boundary conditions combined with nondimensional equations (11)-(15) constitute the mathematical model of two-phase flow in capillary regime with a vapor-liquid phase change. Solution of this mathematical model is not trivial since one of the limit positions remains to be found, the value of L being one of the unknowns of the problem. So, the iteration process is used to solve the mathematical model. To start the calculations, the saturated vapor at the inlet of the tube is assumed having the known mass flux, temperature, pressure, and vapor quality. For the next axial step $\left(z^* + \Delta z^*\right)$, the following steps are executed:

1) To start calculation, an arbitrary total condensation length is assumed;
2) Calculation of the curvature radius from equation (11);
3) Calculation of the local liquid velocity from equation (12);
4) Calculation of the local vapor velocity from equation (13);
5) Calculation of the liquid pressure from equation (14);
6) Calculation of the local vapor pressure from equation (15);
7) Calculation of the total condensation length from equation (26).

Steps 2) to 7) are repeated until the value of the condensation length obtained at the iteration number "it" is approximately equal to the one determined at iteration number "it-1".

Knowing the total condensation length, calculations are then made for the next values of z locations. For each z location, steps 2) to 7) are made and calculation is stopped when z value is approximately equal to the total condensation length L.

5. COMPUTATION OF THE HEAT TRANSFER COEFFICIENT AND THE CONDENSATION LENGTH

5.1. Calculation of the Heat Transfer Coefficient

Assuming that the temperature profile is linear in the condensate film, the local heat transfer is expressed by:

$$h(z) = \frac{\lambda_L}{\delta(z)} \tag{22a}$$

and hence the average heat transfer coefficient in the microchannel condensation length L is performed by the relationship:

$$h_m = \frac{1}{L}\int_0^L h(z)\,\mathrm{d}z \tag{22b}$$

Two last quantities are used to determine the average and the local Nusselt numbers respectively:

$$Nu_{av} = \frac{h_m D}{\lambda_L} \tag{23a}$$

$$Nu_{loc} = \frac{h(z) D}{\lambda_L} \tag{23b}$$

$\delta(z)$ is the average condensate film thickness for each z location along the microchannel. Using the geometrical considerations from the Figure 3, the following analytical expression is established:

$$\delta(z) = \frac{1}{\frac{\pi}{2}}\int_0^{\frac{\pi}{2}} R\left[\frac{\cos(\theta)}{\cos(\Omega)} - 1\right]\mathrm{d}\Omega \tag{24}$$

5.2. Calculation of the Condensate Length

The parameter dR/dz given by the equation (9) is expressed in terms of the limit curvature radius R_L.

At z = L location, dR/dz is to be infinite which causes a calculation problem. To avoid this complication, we set $z = L - \varepsilon$ where ε is an infinitesimal parameter. The limit values of the curvature radius and its first derivative are defined as [22]:

$$R_L = R\big|_{z=L-\varepsilon} = \frac{\varepsilon \sin\beta}{2\cos(\theta+\beta)D_h} \tag{25}$$

$$\left.\frac{\mathrm{d}R}{\mathrm{d}z}\right|_{z=L-\varepsilon} = \frac{\varepsilon\cos\theta - R_L}{R_L\cos\theta} \tag{26}$$

To estimate the numerical value of the condensation length a dichotomy method was performed using equation (26), between the inlet of the microchannels which corresponds to $(z = 0 \text{ and } R = 0)$ and the end of the condensation zone where $(z = L - \varepsilon \text{ and } R = R_L)$.

6. RESULTS AND DISCUSSIONS

6.1. Numerical Results Validation

The validation of the results obtained in the present work for condensation of water in a square microchannel, are compared to the predictions of various correlations available in the literature. These correlations are proposed for condensation heat transfer in microchannels and macrochannels. Among these predictive correlations, those of condensation in microchannels are defined by Wang et al. [26], Koyama et al. [23], and Wang & Rose [14]. Figure 4 shows the comparisons between the average Nusselt number obtained from the present numerical model and those predicted by seven correlations: Dobson et al. [22], Wang et al. [26], Koyama et al. [23], Wang & Rose [14], Traviss et al. [24], Shah [19], and Ackers et al. [25]. Results presented in the Figure 4 are those obtained for condensation of steam in a horizontal square microchannel with hydraulic diameter of 110 μm, for various inlet-

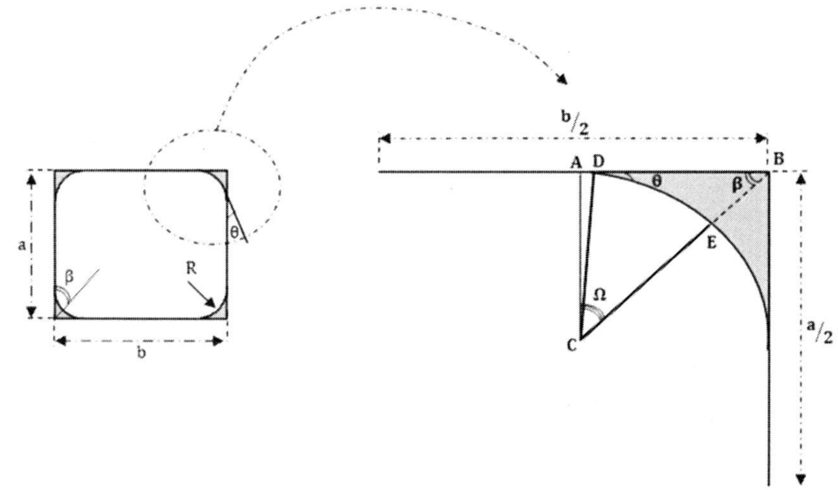

Figure 3. Distribution of the condensate film in the microchannel corners.

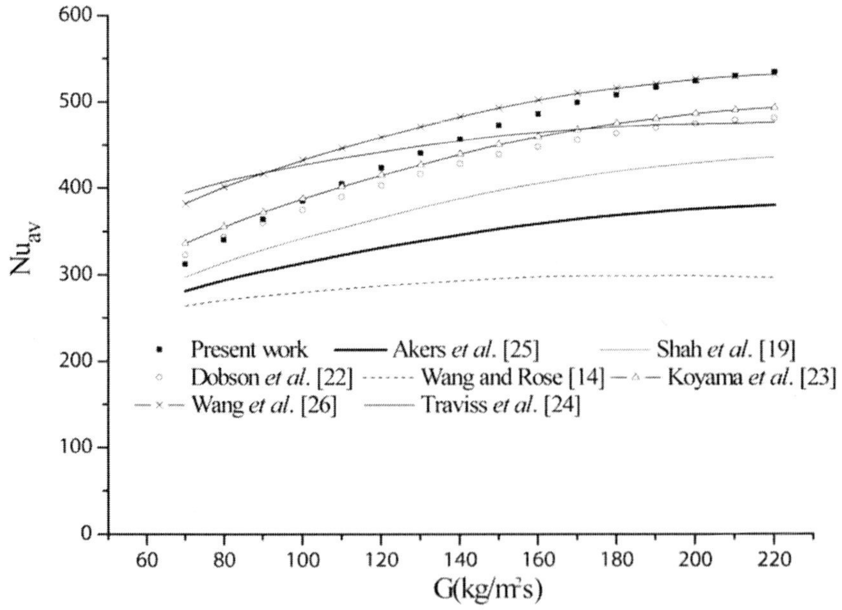

Figure 4. Comparison of the present work with predictions of literature correlations.

vapor mass fluxes ranging from 70 to 220 kg/m^{2}s. It is found that the best predictions of the present average Nusselt number are obtained by the correlations of Dobson et al. [22] and Koyama et al. [23] for which the mean relative deviation is about 7%. Both of these correlations are developed in the case of an annular condensation heat transfer of several pure refrigerants which is carried out in the channels with small hydraulic diameter. The correlations of Traviss et al. [24] and Ackers et al. [25] devoted to the convective condensation heat transfer in the macro-scale channels have been also evaluated to the micro-scale. It can be seen that the correlation of Traviss et al. [24] over predicts highly the present results for low mass fluxes whereas the Ackers's correlation [25] under predicts them highly for all the range of the tested mass fluxes. Adding up, the correlation of Shah [19] is proposed for condensation heat transfer in channels with large hydraulic diameter from 7 to 40 mm and Wang et al. [26] correlation is defined for condensation heat transfer inside a horizontal small rectangular channel with hydraulic diameter of 1.46mm. Average Nusselt numbers obtained by both of these correlations give the same trend and the reasonable results compared to those predicted numerically. The Shah's correlation under predicts numerical results with a maximum average deviation of about 16%. As for the correlation of Wang et al. [26], it over predicts the present results with a maximum deviation of 22% obtained at low mass fluxes. Moreover, correlation of Wang & Rose [14] based on the Nusselt theory including the effects of interfacial shear stress and surface tension on condensation heat transfer is also evaluated. Predictions of Nusselt number from the correlation of Wang & Rose [14] are not in accordance with those of this study since they gave the highest deviation of 37% form the numerical results.

6.2. Peripheral Condensate Film Thickness and Local Heat Transfer

Numerical results are given in the Figures 5(a) and (b) and the Figures 6(a) and (b) for steam condensation in square section microchannel with hydraulic diameter of 110 and 250 μm. Computations were conducted for vapor mass flux G = 90 kg/m^{2}s, contact angle q = 15°, heat flux density q = 100 kW/m^2. For the conditions used in the present computation, the boiling number, the capillary number, the inlet steam temperature and pressure are maintained constant.

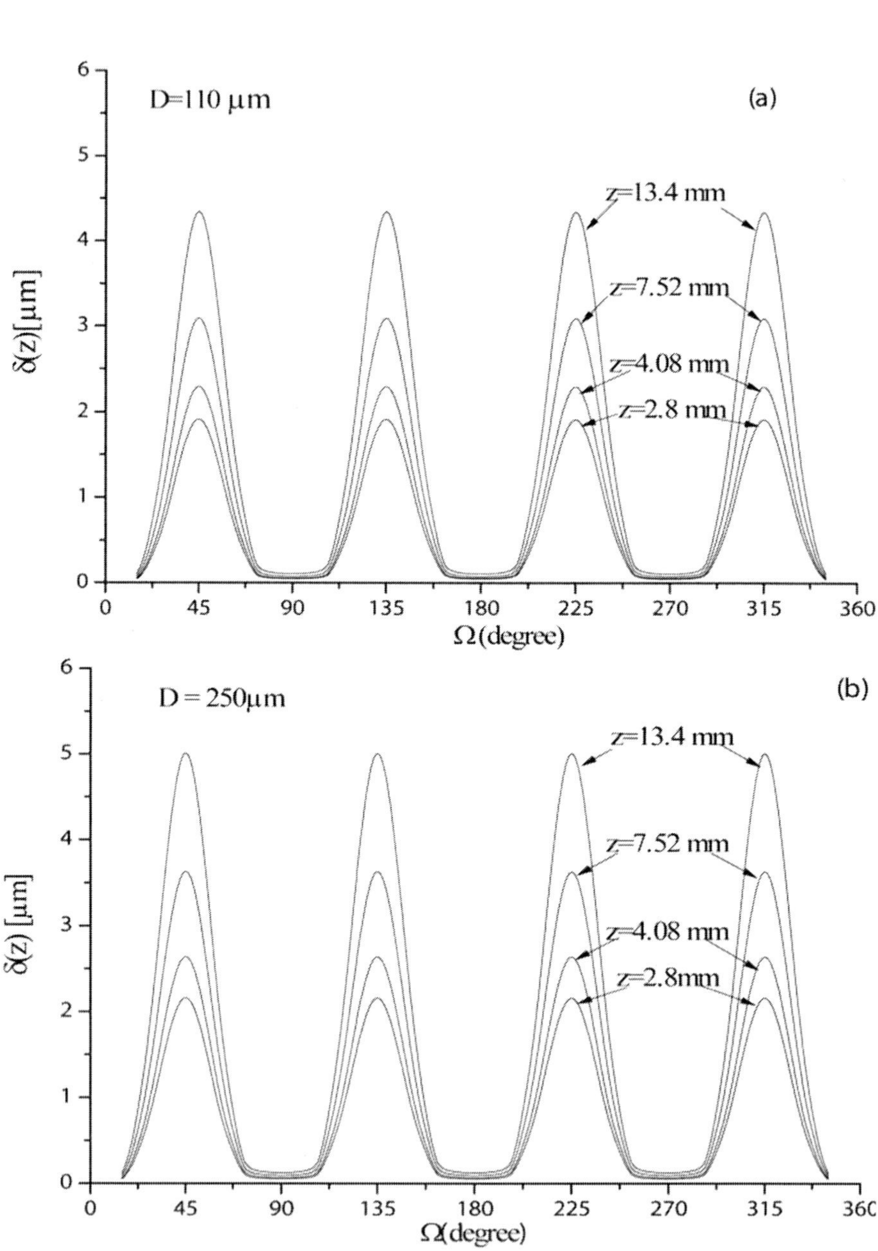

Figure 5. Evolution of the condensate film thickness around the square microchannel circumference.

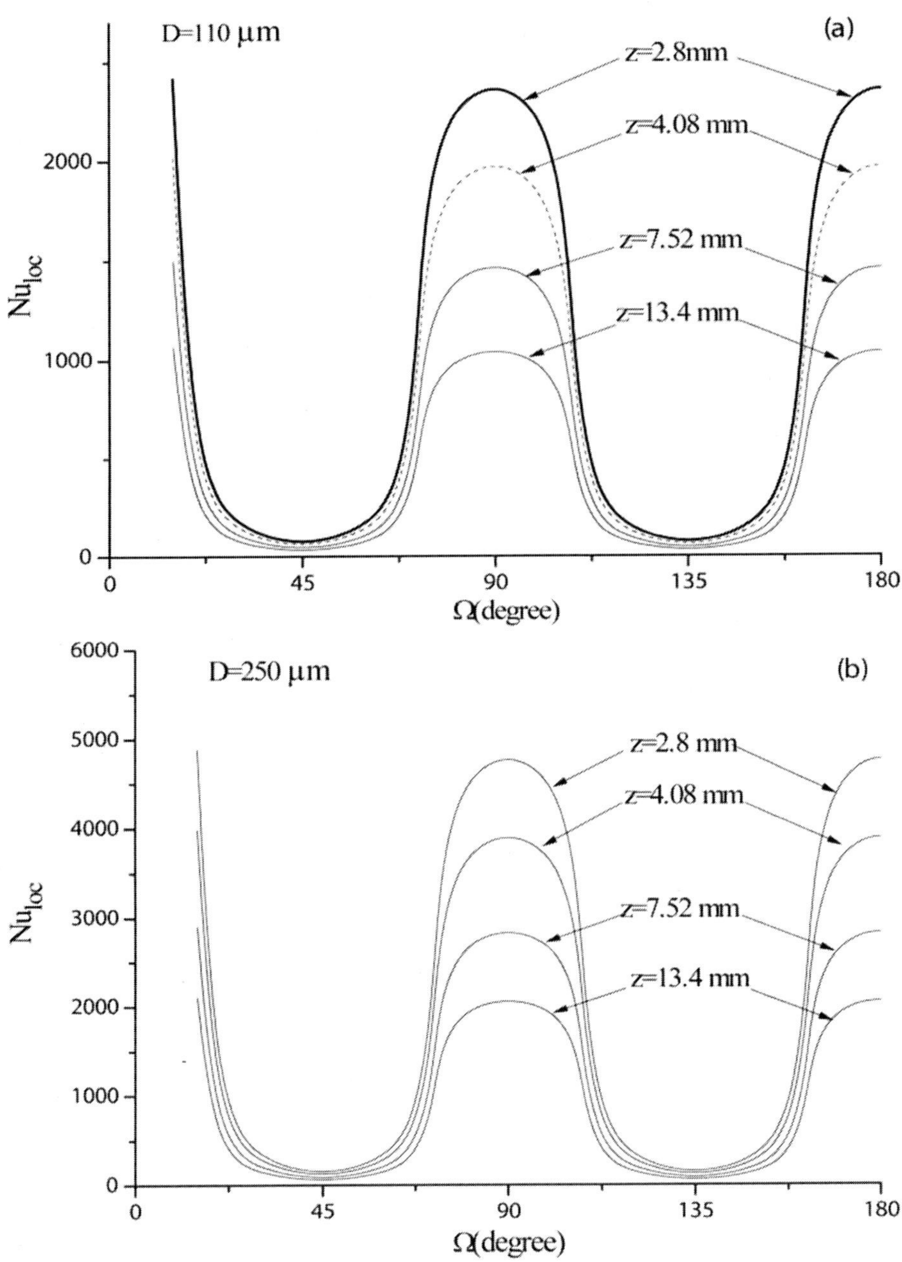

Figure 6. Evolution of the local Nusselt number around the square microchannel circumference.

To better understand the behavior of the peripheral local heat transfer coefficient, Figures 5(a) and (b) show predicted square channel condensate film thickness (24) plotted as a function of the peripheral coordinate W at four different axial locations. It can be seen from these figures that for the same z location on the microchannel, the condensate film thickness is higher for 250 μm (Figure 5(b)) than for 110 μm (Figure 5(a)). According to these representations, it is observed that for both microchannels, the symmetrical distributions of the condensate film thickness are obtained around the channel perimeter because the gravity has no effect on the condensate flow in microchannel. The condensate film thickness is very thick in the microchannel corners (W = 45°, 135°, 225° and 315°) under the effect of the surface tension whereas it becomes very thin at W = 0°, 90°, 180° and 360 where the heat transfer is the highest. These trends have been reversed for the peripheral local heat transfer coefficient (see Figures 6(a) and (b)) even if the symmetrical distributions are conserved.

6.3. Influence of the Cross-Section Shape on the Heat Transfer Coefficient

To indicate the influence of the channel cross-section shape, we start by the presentation of the results concerning curvature radius. In fact, Figure 7 shows the variation of the dimensionless curvature radius along the flow direction in three cross-section shapes of microchannels. At the beginning and the end of the condensation, the curvature radius increases rapidly along the channel, while it increases slowly at the middle. We also note that for the triangular microchannel it's increasing very quickly than other geometries. The present simulation results are similar to those of the reference [21].

The annular condensation length is one of the most important parameter which influences the thermal performance of the microchannel studied here. Figure 8 gives the dimensionless annular condensation length with respect to the boiling number Bo, for three cross-section shapes. As shown in this figure, the condensation zone decreases with Bo and it is also clear that the cross-section shape of the microchannel plays a noticeable role on the condensation.

In the heat transfer exchange point of view, the influence of the microchannel shape on the average Nusselt number is highlighted by studying condensation of steam in a square, equilateral triangular, and rectangular microchannels with the same hydraulic diameter of 250 μm. The sides of the equilateral triangular and square microchannels are 433 μm and 250 μm respectively. For rectangular microchannels having the same hydraulic diameter, the aspect ratio is about 2, 3 and 4. Three different rectangular cross sections are investigated: 375 × 187.5 μm², 500 × 166.6 μm² and 625 × 156.25 μm².

Under the same conditions, Figure 9 compares the average Nusselt numbers with respect to the vapor Reynolds number for five microchannel cross-sections. According to this figure, it is seen that annular condensation Nusselt number is low for square microchannel cross section (aspect ratio of 1). For rectangular cross-section microchannels with aspect ratio (b/a) higher than 1, condensation Nusselt number increases by increasing aspect ratio. The highest Nusselt numbers values are obtained for microchannel with high aspect ratio (equal to 4) and low confinement (a = 156.25 μm). It is interesting to note that for the same hydraulic diameter, the microchannels perimeter and cross section increase with the aspect ratio leading to a reduction of the condensate film thickness. This is due to the predominant effect of the surface tension in the microchannel that increases with microchannel perimeter and thins the condensate film along the sides of the microchannel. For the same Reynolds number, average Nusselt number for equilateral triangular microchannel is between Nusselt number for rectangular microchannel with aspect ratio of 2 and that with aspect ratio of 3. Recall that the perimeter of the triangular microchannel is about 1299 μm and those of rectangular microchannels with aspect ratios of 2 and 3 are about 1125 μm and 1333 μm respectively.

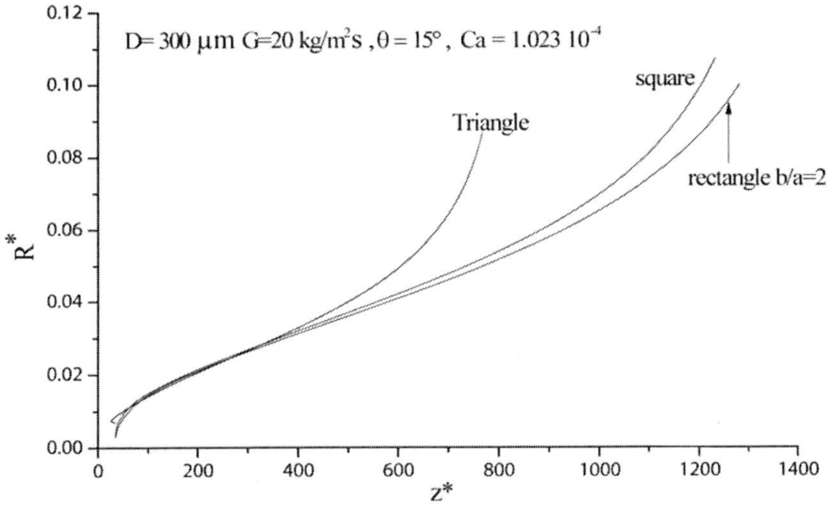

Figure 7. Dimensionless curvature radius along various microchannel shapes.

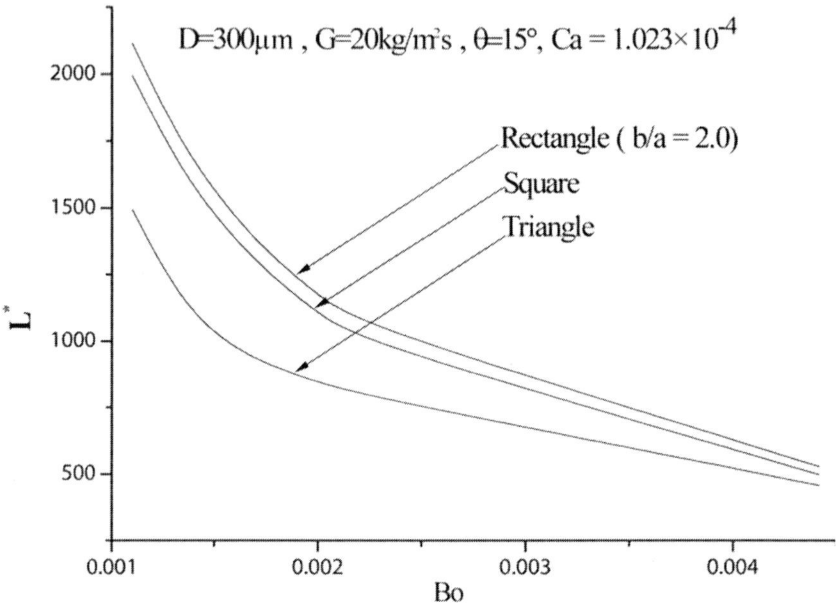

Figure 8. Dimensionless condensation length versus the boiling number.

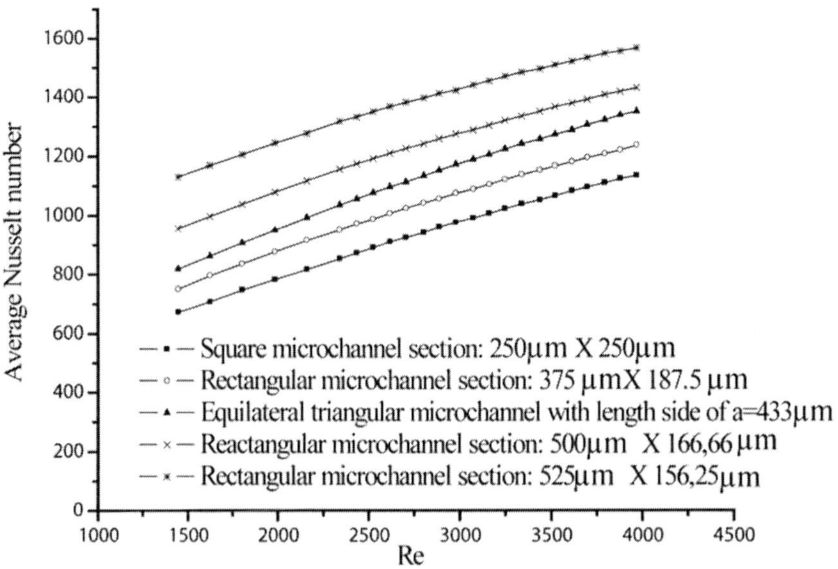

Figure 9. Comparison of the average Nusselt number for various microchannel shapes with the same hydraulic diameter (D = 250μm).

7. CONCLUSION

The numerical model characterizing local heat and mass transfer for condensation in microchannels has been developed by including the effects of wall and liquid vapor interface shear stresses, surface tension, pressure forces, contact angle, etc. Numerical results are compared for steam condensation in square, rectangular, and equilateral triangular microchannels with the same hydraulic diameter. They are compared with the correlations available in the literature and it is shown that correlations for annular condensation in macro-channels predict heat transfer with significant deviations from those obtained for microchannels. The best predictions are obtained with the correlations of Dobson et al. [25] and Koyama et al. [24].

By using the established condensate film thickness expression [24], the influence of the microchannel hydraulic diameter is highlighted. Indeed, increasing this last geometrical factor increases the peripheral condensate film thickness and hence decreases the peripheral condensation heat transfer coefficient. Additionally, we point out that at the beginning and at the end of the condensation, the curvature radius increases rapidly along the channel, while it increases slowly at the middle. We also note that for the triangular microchannel it's increasing quicker than other geometries. As for the annular condensation length, we note that it decreases with increasing of the boiling number and its numerical values diminish in the case of equilateral triangular microchannel.

Finally, to show the major influence of cross-section shape on the condensation heat transfer, comparison of the average Nusselt number is conducted for different microchannel shapes with the same hydraulic diameter. It can be concluded that condensation average heat transfer increases with aspect ratio for rectangular microchannels. The lowest average Nusselt numbers are obtained for the square microchannel because its perimeter and cross section are lower than those of triangular and rectangular microchannels.

REFERENCES

1. S. Kandlikar, S. Garimella, D. Li, S. Colin and M. King, "Heat Transfer and Fluid Flow in Minichannels and Microchannels," Elsevier, Amsterdam, 2006.

2. E. Begg, D. Khrustalev and A. Faghri, "Complete Condensation of Forced Convection Two-Phase Flow in a Miniature Tube," Journal of Heat Transfer, Vol. 121, No. 4, 1999, pp. 904-915.

3. H. Louahlia-Gualous and M. Asbik, "Numerical Modeling of Annular Film Condensation inside a Miniature Tube," Numerical Heat Transfer Part A: Applications, Vol. 52, No. 3, 2007, pp. 251-273.

4. M. Miscevic, P. Lavieille and B. Piaud, "Numerical Study of Convective Flow with Condensation of a Pure Fluid in Capillary Regime," International Journal of Heat and Mass Transfer, Vol. 52, No. 21-22, 2009, pp. 5130-5140.

5. G. B. Ribeiro, J. R. Barbosa and A. T. Prata, "Performance of Microchannel Condensers with Metal Foams on the Air-Side: Application in Small-Scale Refrigeration Systems," Applied Thermal Engineering, Vol. 36, 2012, pp. 152-160.

6. G. El Achkar, P. Lavieille and M. Miscevic, "Loop Heat Pipe and Capillary Pumped Loop Design: About Heat Transfer in the Isolated Bubbles Zone of Condensers," Applied Thermal Engineering, Vol. 33-34, 2012, pp. 253- 257.

7. A. Odaymet, H. Louahlia-Gualous, "Experimental Study of Slug Flow for Condensation in a Single Square Microchannel," Experimental Thermal and Fluid Science, Vol. 38, 2012, pp. 1-13.

8. H. Louahlia-Gualous and B. Mecheri, "Unsteady Steam Condensation Flow Pattern inside a Miniature Tube," Applied Thermal Engineering, Vol. 27, 2007, pp. 1225- 1235.

9. A. Odaymet and H. Louahlia-Gualous, "Experimental Study of Slug Flow for Condensation in a Single Square Microchannel," Experimental Thermal and Fluid Science Vol. 38, 2012, pp. 1-13.

10. X. J. Quan, P. Cheng and H. Y. Wu, "Transition from Annular Flow to Plug/Slug Flow in Condensation of Steam in Microchannels," International Journal of Heat and Mass Transfer, Vol. 51, No. 3-4, 2008, pp. 707-716.

11. T. S. Zhao and Q. Liao, "Theoretical Analysis of Film Condensation Heat Transfer inside Vertical Mini Triangular Channels," International Journal of Heat and Mass Transfer, Vol. 45, 2002, pp. 2829-2842.

12. H. Y. Wu and P. Cheng, "Condensation Flow Patterns in Silicon Microchannels," International Journal of Heat and Mass Transfer, Vol. 48, No. 11, 2005, pp. 2186-2197.

13. H. S. Wang, J. W. Rose and H. Honda, "A Theoretical Model of Film Condensation in Square Section Horizontal Microchannels," Chemical Engineering Research and Design, Vol. 82, No. A4, 2004, pp. 430-434.

14. H. S. Wang and J. W. Rose, "Theory of Heat Transfer during Condensation in Microchannels," International Journal of Heat and Mass Transfer, Vol. 54, No. 11-12, 2011, pp. 2525-2534.

15. A. Agarwal., "Heat Transfer and Pressure Drop during Condensation of Refrigerants in Microchannels," Ph.D. Dissertation, Georgia Institute of Technology, 2006.

16. J. Hu and C. Chao, "An Experimental Study of the Fluid Flow and Heat Transfer Characteristics in Micro-Condensers with Slug-Bubbly Flow," International Journal of Refrigeration, Vol. 30, 2007, pp. 1309-1318.

17. Y. Chen, M. Shi, P. Cheng and G. Peterson, "Condensation in Microchannels," Nanoscale and Microscale Thermophysical Engineering, Vol. 12, 2008, pp. 1-27.

18. Y. Chen, J. Wu, M. Shi and G. Peterson, "Numerical Simulation for Steady Annular Condensation Flow in Triangular Microchannels," International Communications in Heat and Mass Transfer, Vol. 35, No. 7, 2008, pp. 805-809.

19. M. M. Shah, "A General Correlation for Heat Transfer during Film Condensation Inside Pipes," International Journal of Heat and Mass Transfer, Vol. 2, 1979, pp. 547-556.

20. W. M .Kays and R. J. Moffat, "The Behavior of Transpired Turbulent Boundary Layers," In: Studies in Convection: Theory, Measurement and Application, Academic Press, London, 1975.

21. Y. Chen, X. Li, J. Wu and M. Shi, "One Dimensional Numerical Simulation for Steady Annular Condensation Flow in Rectangular Microchannels," Heat Mass Transfer, Vol. 46, No. 1, 2009, pp. 75-82.

22. M. K. Dobson, J. C. Chato, D. K. Hinde and S. P. Wang, "Experimental Evaluation of Internal Condensation of Refrigerants R-12 and R-134a," Proceedings of the ASHRAE Winter Meeting, New Orleans, 23-26 January 1994, pp. 744-754.

23. S. Koyama, K. Kuwahara and K. Nakashit, "Condensation of Refrigerant in a Multi-Port Channel," First International Conference on Microchannels and Minichannels, Rochester, 24-25 April 2003, pp. 193-205.

24. D. P. Traviss and W. M, Rohsenow, "Flow Regimes in Horizontal Two-Phase Flow with Condensation," ASHRAE Transactions, Vol. 79, 1973, pp. 31-39.

25. W. W. Akers, H. A. Deans and O. K. Crosser, "Condensation Heat Transfer within Horizontal Tubes," Chemical Engineering Progress Symposium Series, Vol. 55, No. 29, 1959, pp. 171-176.

26. W. W. Wang, T. D. Radcliff and R. N. A. Christensen, "Condensation Heat Transfer Correlation for MillimeterScale Tubing with Flow Regime Transition," Experimental Thermal and Fluid Science., Vol. 26, No. 5, 2002, pp. 473-485.

6

CHAPTER

THEORETICAL RESEARCH ON R245FA CONDENSATION HEAT TRANSFER INSIDE A HORIZONTAL TUBE

Shengchun Liu*, Yan Huo, Zhaowei Liu, Lan Li, Jinghong Ning

Tianjin Key Laboratory of Refrigeration technology, Tianjin University of Commerce, Tainjin, China

ABSTRACT

Numerical simulation on R245fa condensation inside an inner diameter of 8 mm horizontal tube is researched in this paper. The effect of variation in velocity, condensation temperature and superheat of inlet steam and variation in cooling water temperature on heat transfer coefficient are investigated as a parametric study. Condensation process of steam has been successfully modeled by applying a user defined function (UDF) added to the commercial computational fluid dynamics (CFD) package. By analyzing the corresponding condensate contours and the curves of local heat transfer coefficient, the relationships between condensation heat transfer coefficient and various parameters of R245fa inside horizontal tube are obtained. It shows that the heat transfer coefficient increases by the increase in velocity, condensation temperature and superheat of inlet steam and the decrease in cooling water temperature. The errors between the heat transfer coefficient of simulation result and model of Wang and Shah are within ±30%. The parametric study will provide the basis for designing efficient heat exchangers of R245fa.

Keywords: Condensation Heat Transfer, R245fa, Numerical Simulation, Horizontal Tube

1. INTRODUCTION

Under the dual pressures of energy and environmental issues, developing environmentally friendly refrigerant becomes one of the three major issues facing the refrigeration and air conditioning industry. In recent years, R245fa is an environmentally friendly substituted refrigerant, which is researched by international experts. It has characteristics of excellent thermal resistance, low toxicity, no chlorine, and no damaging effects on the ozone layer. And comparing with R134a, it has a lower pressure level [1] . It is more stable used in some specific situations. As a working fluid of high critical temperature among the zero ODP refrigerants, R245fa represents a substitute for R123 and R11 used in high temperature heat pumps, in jet-pump refrigerators, in centrifugal chillers and in heat recovery power cycles [2] . And R245fa shows high power-generating ability and high economical efficiency in the low-temperature range [3] . R245fa as working fluid in the low-temperature solar power rankine cycle system is feasible and the performance is acceptable [4] . Eames's experiment [5] results demonstrate that low-grade heat can be used to drive efficient jet-pump refrigeration systems, using R245fa as their working fluid, for high temperature applications, such as air conditioning.

It is very important to research the heat transfer characteristics of refrigerants R245fa. Zou et al. [6] used R245fa, R134a, R410A, and R32 to explore the effects of fluid properties on refrigerant distribution in the vertical header of a reversible micro-channel heat exchanger. By comparing the results of experiments, R245fa was found to the best distribution. It is because of the low steam density and high liquid density of R245fa. Many scholars have studied the flow boiling heat transfer of R245fa. In Vakili-Farahani et al.'s experiment [7] with refrigerants R1234ze and R245fa, the heat transfer coefficient increases with rising of saturated temperature, heat flux and mass flux. Tibiriçá et al. [8] did experiment in a horizontal 2.3 mm I.D. stainless steel tube utilizing working fluids R134a and R245fa. It was concluded that the heat transfer coefficient was related to heat flux, mass velocity and quality. Ong et al. [9] , Pike-Wilson et al. [10] and Agostini et al. [11] researched on the flow boiling patterns, pressure drop and heat transfer characteristic of R245fa in a tube. Researching on condensation heat transfer inside tubes is another important aspect of heat transfer characteristics, and the literature on condensation of R245fa is few.

In this paper, condensation heat transfer is simulated by using the commercial software CFD, VOF model and UDF (User-Defined Functions). Numerical simulation on R245fa steam condensing in the inner diameter of 8 mm horizontal tube and analysis is conducted in terms of steam flow rate, cooling water temperature, condensation temperature and superheat. By analyzing the corresponding condensate contours and the curves of local heat transfer coefficient, the relationships between condensation heat transfer coefficient and various parameters inside horizontal tube are obtained. It can provide the basis for designing efficient heat exchangers of R245fa.

2. NUMERICAL METHODOLOGY

A horizontal tube with inner diameter of 8 mm, length 500 mm is considered. And the model coordinates is shown in Figure 1. The origin of the coordinate system is at the center of the interface, the pipe longitudinal direction of the X-axis, the vertical direction for the Z-axis, the influence of gravity is considered in the simulation, and the acceleration of gravity is 9.806 N/kg, the Z-axis is parallel to direction of gravity.

According to pipe geometry, the model is meshed by size of 1 mm using ring hexahedral structured grid. The refrigerant flows into the pipe, and condensation occurs when it comes across the colder wall surface.

The condensation film attached to the wall, and the vapor is in the central area of pipe. Considering the impact of the boundary layer near the wall, Boundary layer near the wall is meshed, and the initial boundary layer mesh size is 0.2 mm, growth ratio of 1.2, the boundary layer mesh layers of 6. The meshing results are shown in Figure 2 and Figure 3.

Figure 4 is a simplified schematic diagram of the simulated section of the tube and boundary conditions, Inlet and outlet boundary condition are set to velocity-inlet and outflow respectively. And constant temperature of the pipe wall is also set.

Many researchers apply VOF model to go on investigation of two phase flow. Such as literature [12] -[14] using it to study boiling and literature [15] [16] using it to research condensation. In this paper, using commercial software Fluent 14.0, multiphase flow model chooses VOF model, calculating as follows within controlling cell.

Within each cell, the volume fraction of phases is 1 for gas-liquid two-phase flow, and as the presence of the following formula:

$$\alpha_l + \alpha_v = 1 \quad (1)$$

Density ρ within each cell, the thermal conductivity λ, the dynamic viscosity μ and thermodynamic energy E are calculated as follows:

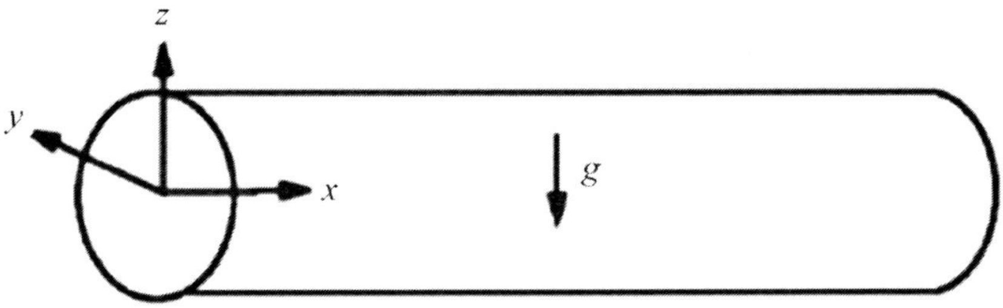

Figure 1. The model coordinates.

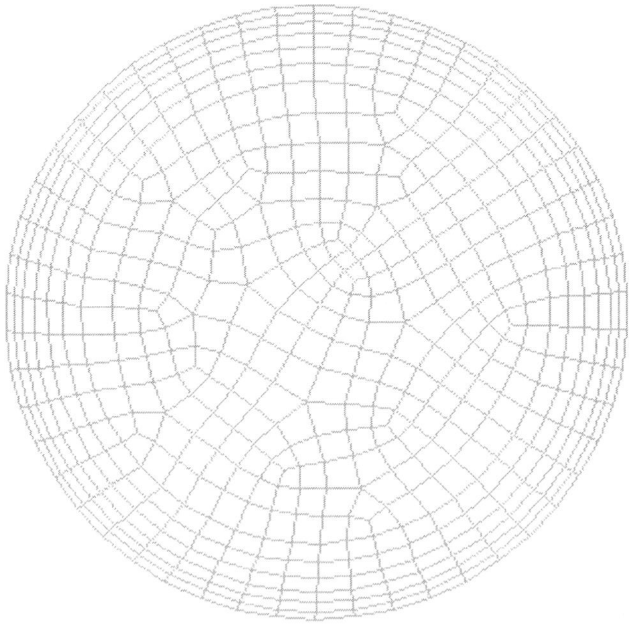

Figure 2. The sectional grid.

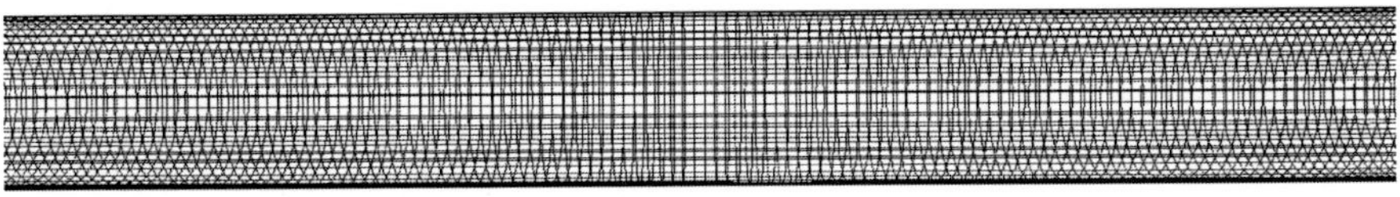

Figure 3. The grid of wall surface.

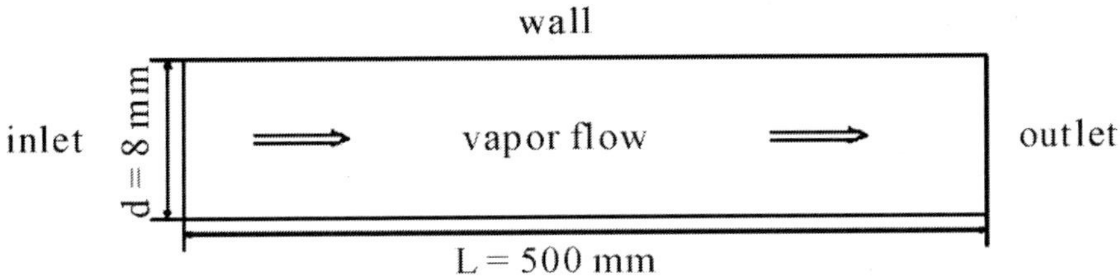

Figure 4. Simplified schematic diagram and boundary conditions.

$$\rho = \alpha_l \rho_l + (1-\alpha_l)\rho_v \tag{2}$$

$$\lambda = \alpha_l \lambda_l + (1-\alpha_l)\lambda_v \tag{3}$$

$$\mu = \alpha_l \mu_l + (1-\alpha_l)\mu_v \tag{4}$$

$$E = \frac{\alpha_l \rho_l E_l + (1-\alpha_l)\rho_V E_v}{\alpha_l \rho_l + (1-\alpha_l)\rho_V} \tag{5}$$

And continuity equation, momentum equation and energy equation can be expressed as follows:

$$\frac{\partial \alpha_l}{\partial \tau} + u \cdot \nabla \alpha_l = \frac{S_l}{\rho_l} \tag{6}$$

$$\frac{\partial \alpha_v}{\partial \tau} + u \cdot \nabla \alpha_v = \frac{S_v}{\rho_v} \tag{7}$$

$$\frac{\partial(\rho u)}{\partial \tau} + \nabla \cdot (\rho u u) = -\nabla p + \nabla \cdot \left[\mu\left(\nabla u + \nabla u^T\right)\right] + \rho g + F \tag{8}$$

$$\frac{\partial(\rho E)}{\partial \tau}+\nabla\cdot\left[u(\rho E+p)\right]=\nabla\cdot\left(\lambda_{eff}\nabla T\right)+Q \tag{9}$$

In the present CFD model, the liquid evaporation and vapor condensation effects can be considered by incorporating mass and energy sources into the continuity and energy equations, respectively. The source term for the energy equation can be obtained by multiplying the rate of mass transfer by the latent heat. In a CFD simulation, the source terms act on the whole fluid domain, so the phase change effect both at liquid surface and in the bulk region can be considered simultaneously. All of the source terms are driven by the cell temperature, T, and implemented via customized UDF. Based on the following temperature regimes, the mass transfer can be described as follows [17] :

If $T \geq T_{sat}$ (evaporation),

$$m=\varphi\alpha_l\rho_l\frac{|T-T_{sat}|}{T_{sat}} \tag{10}$$

If $T < T_{sat}$ (condensation),

$$m=\varphi\alpha_v\rho_v\frac{|T-T_{sat}|}{T_{sat}} \tag{11}$$

Using the k-ε turbulence model, the formats of simulation variables to select are shown in Table 1, and the convergence criterion of control equations are shown in Table 2.

3. RESULTS AND DISCUSSION

Refrigerants R245fa inlet steam is saturated, the condensation temperature is 314 K, the wall temperature is 293 K, and the inlet steam flow rate is 6 m/s, 10 m/s, 14 m/s and 18 m/s respectively. The changes of film thickness of the inner tube and heat transfer coefficient are shown in Figure 5.

From above curves, it is concluded that the local heat transfer coefficient increased, the thickness of condensate film decreased and the rate of increase of condensation film thickness slow down, as the inlet steam flow

Table 1. Format of variables.

Variables	Format
Solver	Pressure based
Time	Steady
VOF format	Implicit
Pressure-velocity coupling	SIMPLE
Pressure	Body force weighted
Momentum	Second-order upwind
Volume fraction	First-order upwind
Turbulent kinetic energy	Second-order upwind
Turbulent dissipation rate	Second-order upwind
Energy	Second-order upwind

Table 2. The convergence criterion of control equations.

Equations	Residuals
Continuity	1×10^{-3}
Momentum	1×10^{-3}
Energy	1×10^{-3}
Turbulence	1×10^{-3}

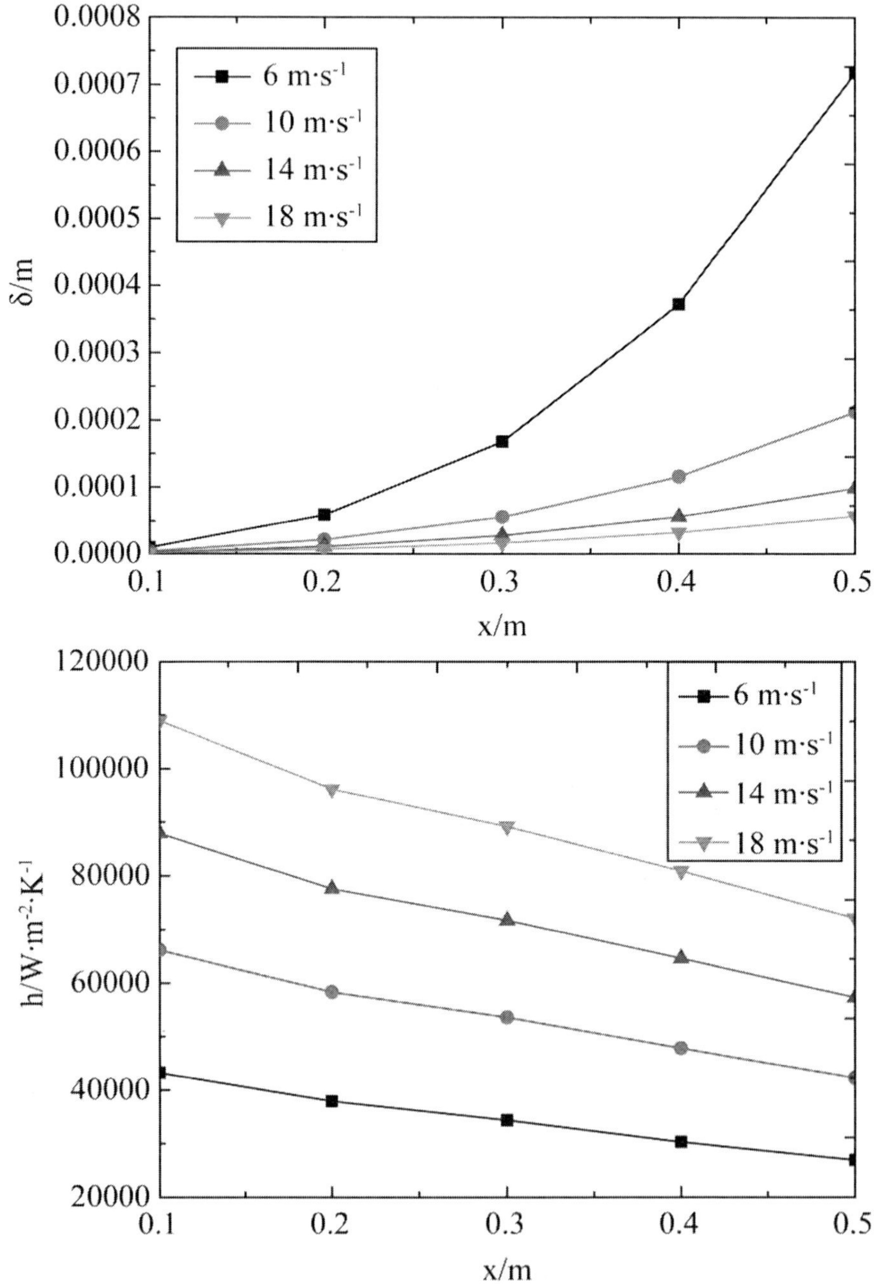

Figure 5. Curves under different flow rates.

rate increases. And condensation heat transfer coefficient changes dramatically, when refrigerant steam flows into tube. At the Steam flow rate of 6 m/s, condensate begins to increase rapidly in 0.3 m at growth gradually accelerated. As the flow rate increases, the changing rate of local heat transfer coefficient increases. The faster steam flow rate, the stronger the shear is, the more slowly the condensation film thickness increases.

The inlet velocity is 10 m/s, and the contours of liquid volume fraction are shown in Figure 6 and Figure 7, respectively. The surface heat transfer coefficient is shown in Figure 8. The surface heat transfer coefficient increases at first and then decreases, therefore there is a maximum, and the rate of increase becomes faster close to the entrance of steam. It can be interpret as part of the steam condensed to form droplets at the inlet; with the process of condensation, droplets become larger, increasing the heat transfer area. Then the droplets become liquid film, film formation increases heat resistance, making the heat transfer coefficient decreases.

As shown in Figure 7 and Figure 8, because the steam temperature is higher than that of pipe wall, the steam condenses on the wall. With the working fluid flowing and the occurrence of condensation, the film thickness increased along direction of flowing. Due to gravity, most of liquid accumulates in the bottom of the pipe. The condensation heat transfer coefficient changed dramatically as refrigerant steam flowed into tube. Then, the heat transfer coefficient reduced, the heat transfer coefficient is further reduced quickly at the position of about 0.3 m.

Refrigerants R245fa inlet steam is saturated, the inlet steam flow rate is 10 m/s, the condensation temperature is 314 K, and the wall temperature was 288 K, 291 K, 293 K and 295 K respectively. The changes of film thickness of the inner tube and heat transfer coefficient are shown in Figure 9. The lower the temperature of the tube wall, the great heat transfer coefficient is, the more condensed liquid is. Analysis of the causes is that the lower tube wall temperature, the greater heat temperature difference, and the larger condensation heat transfer. Condensate begins to increase rapidly at 0.3 m and the thickening rate of film increased, heat transfer coefficient decreases rapidly.

Refrigerants R245fa inlet steam is saturated, the inlet steam flow rate is 10 m/s, the wall temperature is 293 K, and the condensation temperature is 308 K, 313 K and 318 K, respectively. The changes of film thickness of the inner tube and heat transfer coefficient are shown in Figure 10.

Changes of the film thickness and heat transfer coefficient along tube are similar to the cooling water temperature trends under condensation temperatures. The larger the heat transfer coefficient, the thicker the condensation film. The rate of condensation film changing become fast with the rising of condensation temperature. These can be expressed as the higher condensation temperature, the greater the temperature difference, there is a growing rate of condensation. That condensate begins to increase rapidly at 0.3 m and the rate of increase gradually accelerated can also be obtained.

Refrigerants R245fa inlet steam is superheated with 1 K, 5 K, 10 K and 15 K. The inlet steam flow rate is 10 m/s, the wall temperature is 293 K, and the condensation temperature is 314 K, respectively. The changes of film thickness and heat transfer coefficient are shown in Figure 11.

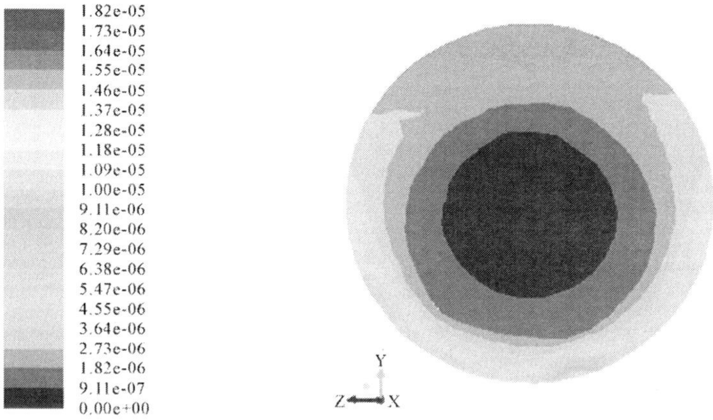

x = 0.2 m

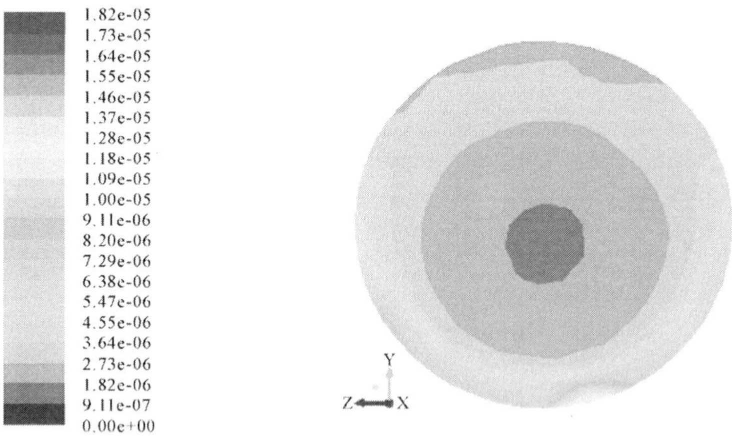

x = 0.3 m

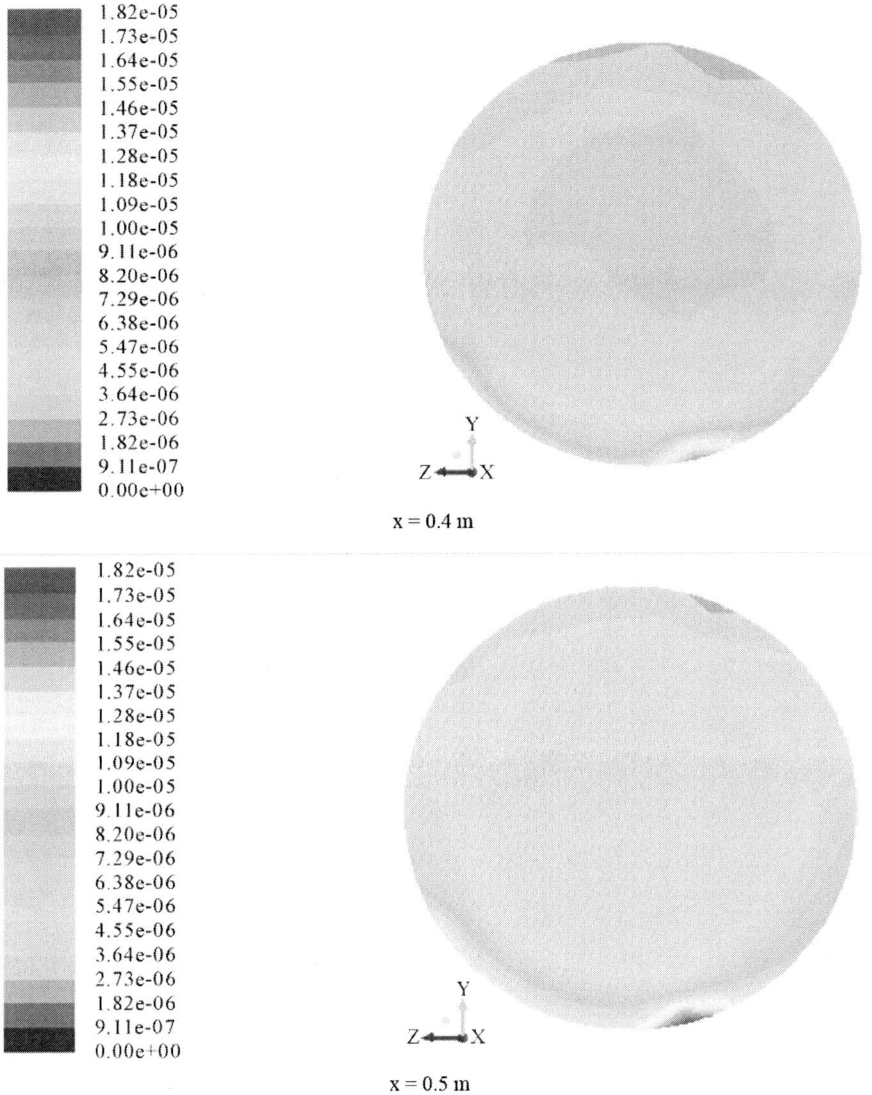

Figure 6. The contours of liquid volume fraction at different positions.

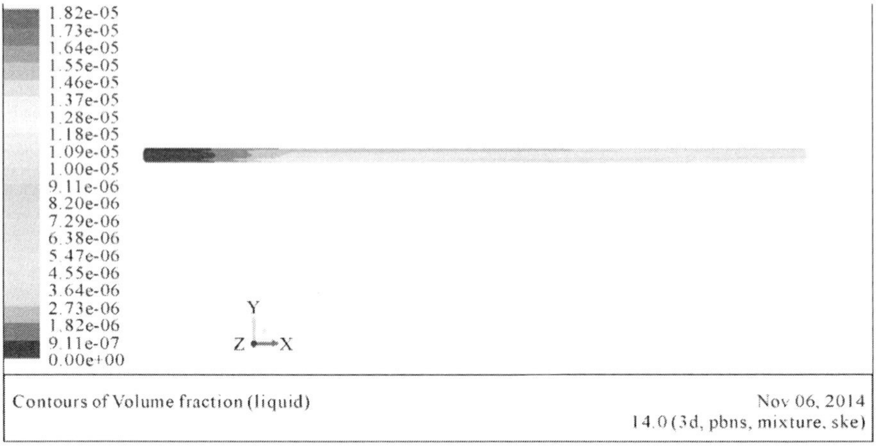

Figure 7. The contours of liquid volume fraction along the tube at the velocity of 10 m/s.

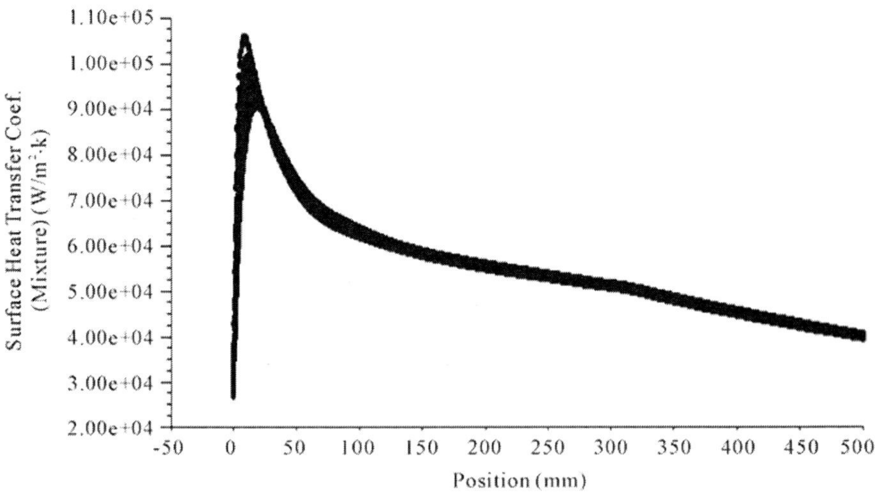

Figure 8. The surface heat transfer coefficient along the tube at the velocity of 10 m/s.

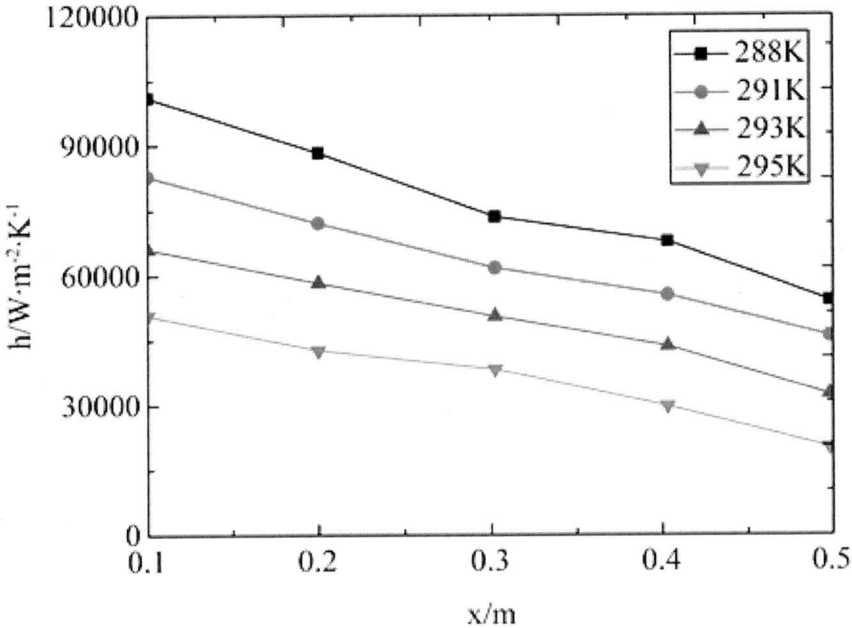

Figure 9. Curves under different wall temperature.

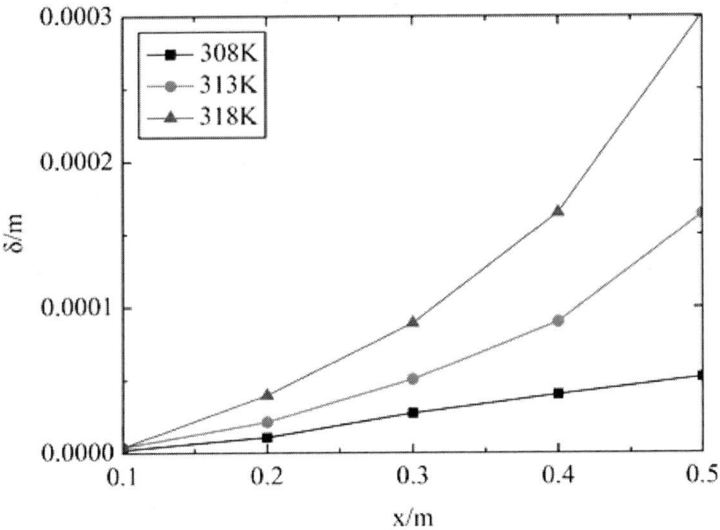

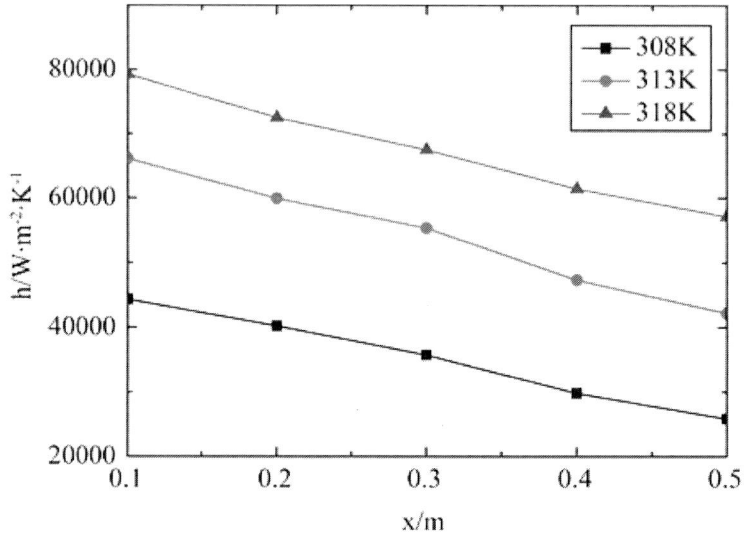

Figure 10. Curves under different condensation temperature.

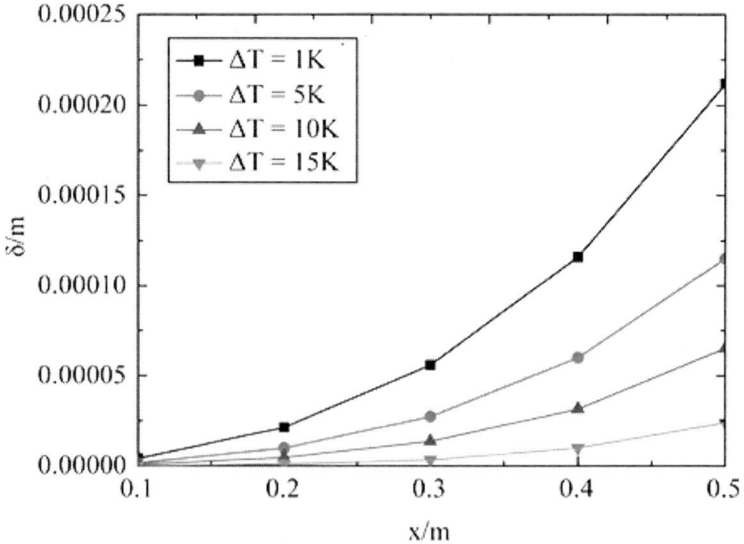

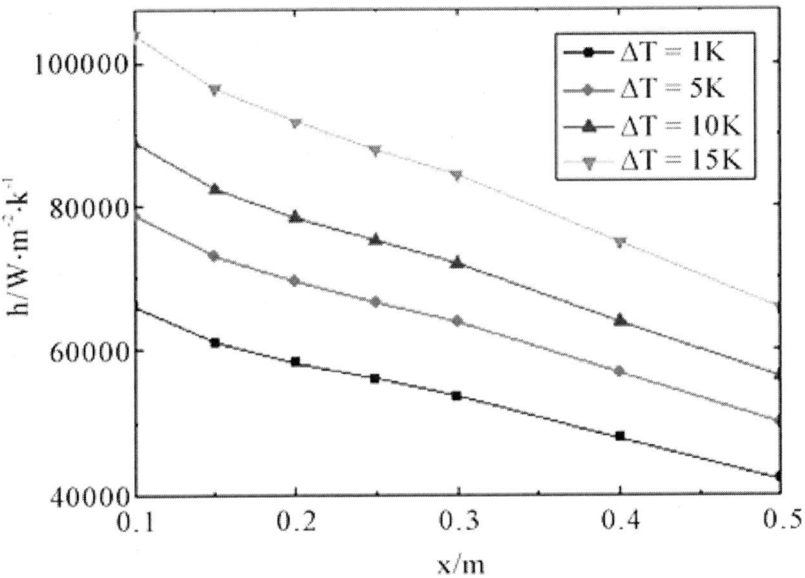

Figure 11. Curves under different superheat.

When the superheat increases, the heat transfer coefficient becomes larger and the amount of condensate reduces. Superheat temperature difference leads to heat transfer coefficient increasing, but the amount needed for the cold steam condensation increases. Film thickness starts to increase rapidly at 0.3 m, the growth rate increases gradually, and the curve of corresponding heat transfer coefficient has the tendency to decline rapidly.

As the inlet steam flow rateis 10 m/s, the wall temperature is 293 K, and condensation temperature is 308 K, the errors between the heat transfer coefficient of simulation result and model of Wang [18] and Shah [19] are within ±30%.

4. CONCLUSIONS

By simulation on the process of R245fa condensing inside a horizontal tube and analysis of simulation results, various parameters have great influences on the condensation heat transfer coefficient. The simulation results show that when the inner diameter of tube is 8 mm, the heat transfer coefficient and film thickness have significant changes after the position of 0.3 m.

Considering the changes of film thickness and heat transfer coefficient under different conditions, some relevant measures can be taken to reduce the film thickness at about 0.4 m and to enhance heat transfer in this case. The changing regulars of film thickness and heat transfer coefficient in different diameter can be got by using the similar method of simulation in the paper. Then, the position can be obtained to enhance heat transfer. And it can provide the basis for designing efficient heat exchangers of R245fa.

By comparing the different factors on the condensation heat transfer, it can be seen that the ways of influences of cooling water temperature, condensing temperature and steam flow rate on the condensation heat transfer are different. The increasing of team flow rate results in enhanced shear, and condensation film is not stable in the presence of the tube, so film thickness reduces, and heat transfer coefficient increases. The wall temperature decreases or condensing temperature rises, and the temperature difference increases, so that the heat transfer coefficient increases. The heat transfer coefficient of simulation results is close to the model of Wang and Shah. Thus, the model and method of simulation are suitable for process of condensation.

ACKNOWLEDGEMENTS

This research was supported by The Ministry of Science and Technology of the People's Republic of China, (863 Program) (Grant No. 2012AA053001).

REFERENCES

1. Wang, X.D., Zhao, L. and Wang, J.L. (2000) Experimental Investigation on the Low-Temperature Solar Rankine Cycle System Using R245fa. Energy Conversion and Management, 52, 946-952.
2. Angelino, G. and Invernizzi, C. (2003) Experimental Investigation on the Thermal Stability of Some New Zero ODP Refrigerants. International Journal of Refrigeration, 26, 51-58.
3. Saleh, B., Koglbauer, G., Wendland, M. and Fischer, J. (2007) Working Fluids for Low-Temperature Organic Rankine Cycles. Energy, 32, 1210-1221.

4. Wang, X.D., Zhao, L., Wang, J.L., Zhang, W.Z., Zhao, X.Z. and Wu, W. (2010) Performance Evaluation of a Low- Temperature Solar Rankine Cycle System Utilizing R245fa. Solar Energy, 84, 353-364.

5. Eames, I.W., Ablwaifa, A.E. and Petrenko, V. (2007) Results of an Experimental Study of an Advanced Jet-Pump Refrigerator Operating with R245fa. Applied Thermal Engineering, 27, 2833-2840.

6. Zou, Y. and Hrnjak, P.S. (2014) Effects of Fluid Properties on Two-Phase Flow and Refrigerant Distribution in the Vertical Header of a Reversible Microchannel Heat Exchanger—Comparing R245fa, R134a, R410A, and R32. Applied Thermal Engineering, 70, 966-976.

7. Vakili-Farahani, F., Agostini, B. and Thome, J.R. (2013) Experimental Study on Flow Boiling Heat Transfer of Multiport Tubes with R245fa and R1234ze(E). International Journal of Refrigeration, 36, 335-352.

8. Tibiriçá, C.B. and Ribatski, G. (2010) Flow Boiling Heat Transfer of R134a and R245fa in a 2.3 mm Tube. International Journal of Heat and Mass Transfer, 53, 2459-2468.

9. Ong, C.L. and Thome, J.R. (2009) Flow Boiling Heat Transfer of R134a, R236fa and R245fa in a Horizontal 1.030 mm Circular Channel. Experimental Thermal and Fluid Science, 33, 651-663.

10. Pike-Wilson, E.A. and Karayiannis, T.G. (2014) Flow Boiling of R245fa in 1.1 mm Diameter Stainless Steel, Brass and Copper Tubes. Experimental Thermal and Fluid Science, 59, 166-183.

11. Agostini, B., Thome, J.R., Fabbri, M., Michel, B., Calmi, D. and Kloter, U. (2008) High Heat Flux Flow Boiling in Silicon Multi-Microchannels—Part II: Heat Transfer Characteristics of Refrigerant R245fa. International Journal of Heat and Mass Transfer, 51, 5415-5425.

12. Rabha, S.S. and Buwa, V.V. (2009) Volume-of-Fluid (VOF) Simulations of Rise of Single/Multiple Bubbles in Sheared Liquids. Chemical Engineering Science, 65, 527-537.

13. Yang, Z., Peng, X.F. and Ye, P. (2008) Numerical and Experimental Investigation of Two Phase Flow during Boiling in a Coiled Tube. International Journal of Heat and Mass Transfer, 51, 1003-1016.

14. Huang, M., Yang, Z., Duan, Y.Y. and Lee, D.J. (2011) Bubble Growth for Boiling Bubbly Flow for R141b in a Serpentine Tube. Journal of the Taiwan Institute of Chemical Engineers, 42, 727-734.

15. Chen, S.H., Yang, Z., Duan, Y.Y., Chen, Y. and Wu, D. (2014) Simulation of Condensation Flow in a Rectangular Microchannel. Chemical Engineering and Processing, 76, 60-69.

16. Da Riva, E. and Del Col, D. (2012) Numerical Simulation of Laminar Liquid Film Condensation in a Horizontal Circular Minichannel. Journal of Heat Transfer, 134, Article ID: 051019.

17. Wang, L., Li, Y.Z., Li, C. and Zhao, Z.X. (2013) CFD Investigation of Thermal and Pressurization Performance in LH2 Tank during Discharge. Cryogenics, 57, 63-73.

18. Wang, W.W.W., Radcliff, T.D. and Christensen, R.N. (2002) A Condensation Heat Transfer Correlation for Millimeter-Scale Tubing with Flow Regime Transition. Experimental Thermal and Fluid Science, 26, 473-485.

19. Shah, M.M. (2010) Heat Transfer during Condensation inside Small Channels: Applicability of General Correlation for Macrochannels. Proceedings of the 14th International Heat Transfer Conference, Washington DC, 8-13 August 2010, 298-303.

7

CHAPTER

EXPERIMENTAL ANALYSIS OF HEAT TRANSFER BEHAVIOR INSIDE HEAT PIPE INTEGRATED WITH COOLING PLATES

Chen-Ching Ting[1*], Chien-Chih Chen[2]

[1]*Department of Mechanical Engineering, National Taipei University of Technology, Chinese Taipei*

[2]*Institute of Mechanical and Electrical Engineering, National Taipei University of Technology, Chinese Taipei*

ABSTRACT

This work used experimental methods to study heat transfer behavior inside a heat pipe and found that heat transfer behavior inside the heat pipe was changed due to its integration with cooling plates. This change caused the heat pipe to have copper-like heat transfer behavior. Experimental performances first built a CPU simulator with maximum heat power 300 W in accordance with the ASTM standard as heat source and measured temperature distribution by using infrared thermography and thermocouple thermometer. Observation of heat transfer behavior inside heat pipe influenced by its integration with cooling plates used color schlieren technique. A commercial CPU heat pipe cooler was also used as reference object in this work. Integration of the heat pipe with cooling plates causes the heat pipe to have the copper-like heat transfer behavior. The results indicate that rebuilding the bare heat pipe's heat transfer behavior is the best solution for improving cooling efficiency of the heat pipe cooler.

Keywords: Heat Transfer Behavior, Heat Pipe, Color Schlieren, Infrared Thermography, Heat Pipe Cooler

1. INTRODUCTION

The general principle of heat pipe employs evaporative cooling on the heat source with the working fluid and its capillary action transports the condensed fluid back to the evaporating section. It was first noted by Grover at Los Alamos National Laboratory in 1963 and subsequently published in the Journal of Applied Physics in 1964 [1]. A heat pipe is a passive cooling device with extremely high thermal conductivity ca. 5000 ~ 30,000 [W/mK], which is initiated by a special heat transfer mechanism and can transport relatively large amount of heat in a tiny temperature difference between the hot and the cold interfaces. The heat pipe is hence often named as the thermal superconductor. Stenger in 1966 [2] first developed the technique of capillary pumped loop (CPL) in heat pipe and Maydanik et al. in 1985 originally developed the loop heat pipe (LHP) for the terrestrial solar energy market using the technique of anti-gravitation heat pipe [3].

The heat transfer property of heat pipe is theoretically different from that of copper pipe. Heat pipe is spatially discrete, whereas the copper pipe is continuous. Heat pipe for cooling has been widely used in the market due to its extremely large value of thermal conductivity and normally combined with the cooling finny pieces, named as the heat pipe cooler. The heat pipe cooler should have theoretically excellent cooling efficiency, but the fact is that it cannot reach the predicted value. This is the main question for applications of the heat pipe cooler. Most of the studies on the heat pipe cooler ignored the considerations on the heat transfer behavior change of heat pipe due to the integration with the cooling plates and discussed only its integrating structure. Bejan and Scibba in 1992 [4] indicated that the optimal condition of the cooling plates is dependent on the three parameters, $L^{1/2}$, $(\mu\alpha)^{1/2}$, $(\Delta P)^{-(1/4)}$,for the heat pipe cooler, where L is length of the cooling plates, μ is viscosity of the cooling fluid, α is thermal diffusivity of the cooling fluid, and ΔP is pressure difference of the in and out cooling fluid. Yeh and Chang in 1995 [5] show an optimal design for heat pipe cooler. There are a lot of studies about the improvement of the cooling efficiency for heat pipe cooler, but their discussions were almost focused on the integrating structural design [6-8].

Studies on the heat transfer behavior inside the heat pipe are relatively few. A standard application of heat pipe is in connection with cooling plates, but the integration has changed the original heat transfer property of the heat pipe. The heat transfer behavior of heat pipe is similar to copper pipe due to the integration with cooling plates, which certainly reduces its cooling efficiency.

This work discusses the inner heat transfer property change of heat pipe integrated with cooling plates using experimental methods. It found that the change of heat transfer behavior inside heat pipe integrated with cooling plates is the main reason for the lack cooling efficiency. The cooling plates far from the heat source cannot reach sufficient cooling efficiency. This work found also that a new structure of heat pipe for integration with cooling plates should be designed to keep the original heat transfer behavior of bare heat pipe after integration with cooling plates [9].

2. BASIC THEORY

A heat pipe is theoretically divided into three parts, the evaporator, the adiabatic section, and the condenser. Figure 1 shows schematics of a bare heat pipe. In process, the working fluid absorbs heat from heat source and evaporates in the evaporator. The hot vapor directly runs through the adiabatic section and condenses in the condenser. The condensed working fluid in the condenser runs back by way of the capillary structure on the inner wall in the adiabatic section and reaches the evaporator again.

Figure 2 shows schematics of a common heat pipe cooler, where the convective heat transfer to the cooling fin continuously decreases. A common heat pipe cooler is integrated with cooling fin in the adiabatic section. Due to integration of the cooling fin with heat pipe in the adiabatic section, the heat of the vapor is forcedly taken away by the cooling fin in the adiabatic section and therefore condenses earlier. In other words, the vapor cannot reach the condenser section and the adiabatic section disappears. This result will shorten the effective cooling length of the heat pipe. The original heat transfer property of the heat pipe integrated with cooling fin in the adiabatic section is unfortunately destroyed by integration of the cooling fin with the heat pipe and is changed to be similar to copper pipe, where the bare heat pipe is discrete and the copper pipe is continuous.

To increase cooling efficiency of a heat pipe cooler, temperature distribution on the heat pipe should be more uniform. That is, the original heat transfer property of the heat pipe in a heat pipe cooler should be recovered. In other words, the adiabatic section must be kept. A future application of this work is the so-called dual-pipe heat pipe cooler [10]. Figure 3 shows schematics of a dual-pipe heat pipe cooler, where the convective heat transfer to the cooling fin is more uniform. A dual-pipe heat pipe builds the inner coaxial pipe in the adiabatic section of a common heat pipe. The built inner coaxial pipe can avoid heat of the vapor to be taken away in the adiabatic section. This special design lets the vapor be able to run through the adiabatic section and successfully recover the original heat transfer property of the heat pipe for a heat pipe cooler.

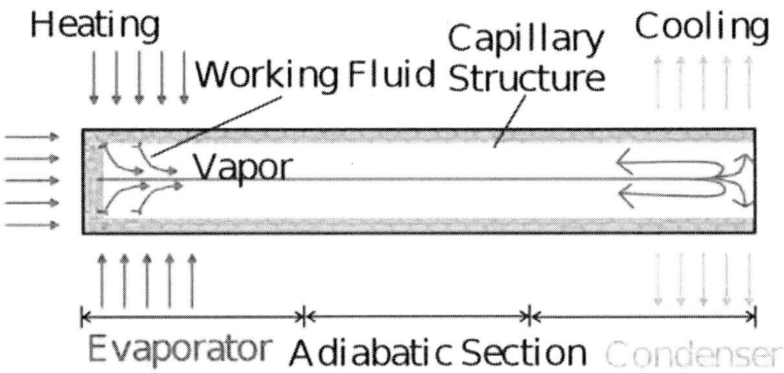

Figure 1. Schematics of a bare heat pipe [10].

3. EXPERIMENTAL DETAILS

A CPU simulator was built as the heat source in this work and the experiments were done by using the color schlieren, the infrared thermography, and the thermocouple thermometer for temperature determination. The built CPU simulator in accordance with the ASTM D5470 standard consists of a T form copper block, some glass wool, fireproof board, two 150 W cartridge heaters, and a 300 W power supply. Figure 4 shows the schematic description of the built CPU simulator, where the temperature on the three points T_u,, T_m, and T_l are used to determine the simulated CPU power.

According to the Fourier's equation:

$$\dot{Q} = -kA\Delta T \tag{1}$$

where $\dot{Q}$ is the rate of heat flow over the area A and T is temperature. The negative sign indicates that the temperature gradient is in the opposite direction to the heat flow and k is the thermal conductivity of the material. Application of the one-dimensional Fourier's equation for determination of the CPU simulator output heat flow rate $\dot{Q}_{out}$ receives

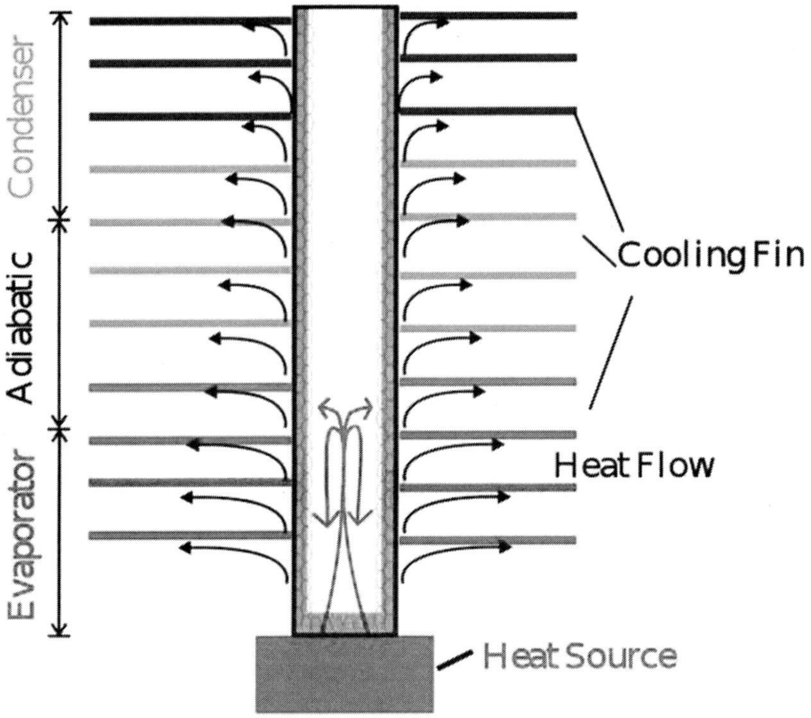

Figure 2. Schematics of a heat pipe cooler [10].

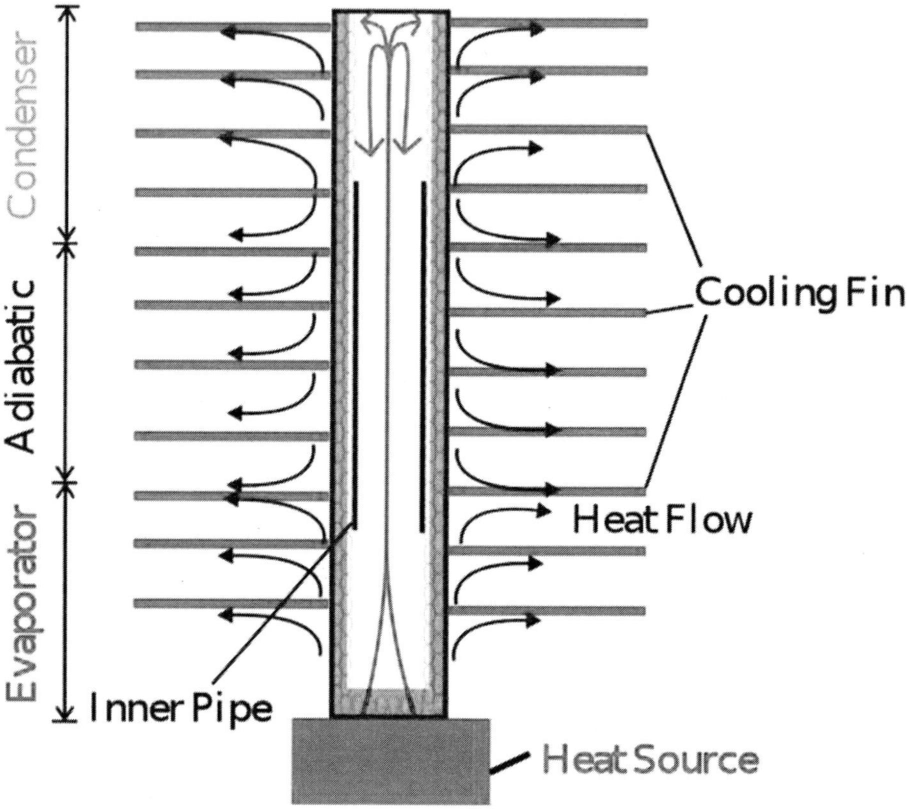

Figure 3. Schematics of a dual-pipe heat pipe cooler [10].

$$\dot{Q}_{out} = kA\frac{T_l - T_u}{\Delta x} \tag{2}$$

where Δx is the positional difference of points T_l and T_u, A is 31 × 31 mm^2. Figure 5 shows the calibration curve of the input and output powers corresponding to the power supply and the CPU simulator, respectively. The function of the fitted curve is y = 0.93x – 0.31 with the fitting error smaller than 2%.

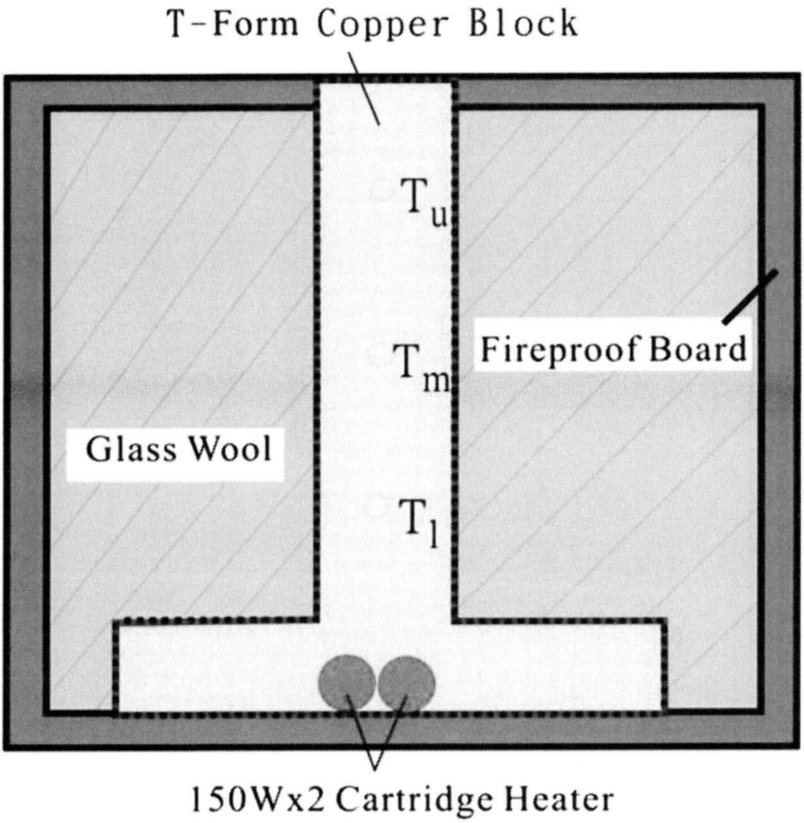

Figure 4. Schema of the CPU simulator's side cross section.

Figure 6 shows photo of the heat pipe built on the CPU simulator. The experimental setup used the IRISYS infrared thermal camera, which has resolution of 16 × 16 pixels and capturing speed of 8 fps, to record the time dependent change of the temperature distribution on the heated heat pipe.

The capturing screen of the infrared thermal camera shows the 16 × 16 pixels corresponding to 16 × 16 temperature sensors. Figure 7 represents the corresponding positions with the heated heat pipe, which will be applied to the data evaluation.

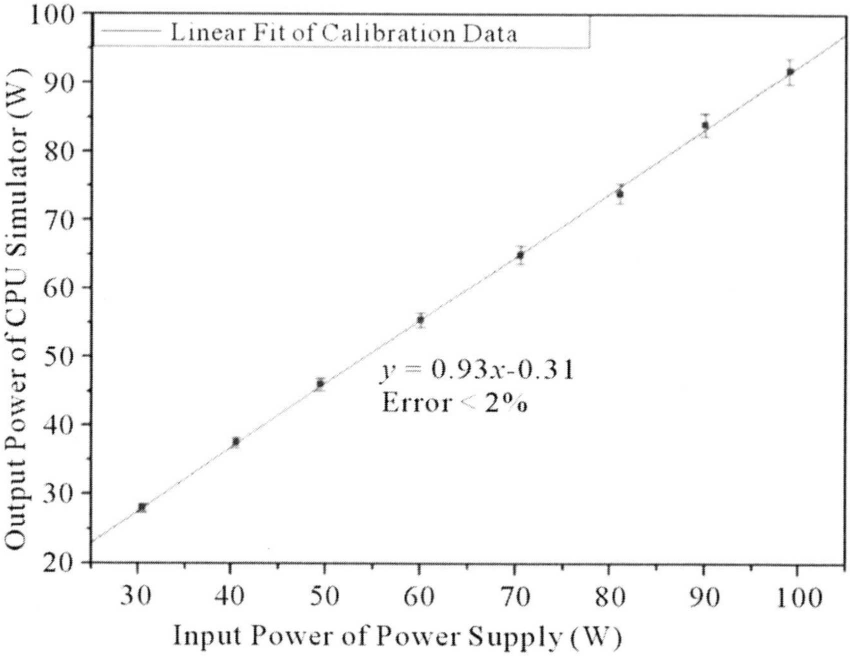

Figure 5. Calibration curve of the CPU simulator [9].

Observation of the heat transfer behavior inside the heat pipe influenced by the integration with cooling plates used the color schlieren technique. This experimental setup was built in water tank due to the water has larger thermal conductivity than air. The water was set as a good cooling medium for the heat pipe. Figure 8 shows the schematic setup of the heat pipe integrated with cooling plates. The sequent heating by the heat source convects heat into the water and yields the simultaneous heat flux which was visualized by the color schlieren technique.

A commercial heat pipe cooler with U-form heat pipe was used as the reference object. Its heat pipe has total length 265 mm and diameter ϕ 6 mm in connection with 25 pieces cooling plates which have each thickness 0.46 mm and area 87 × 20 mm^2. There are 5 different points on the heat pipe for determination of the temperature distribution. Figure 9 shows the sketch and Figure 10 is photo of the setup.

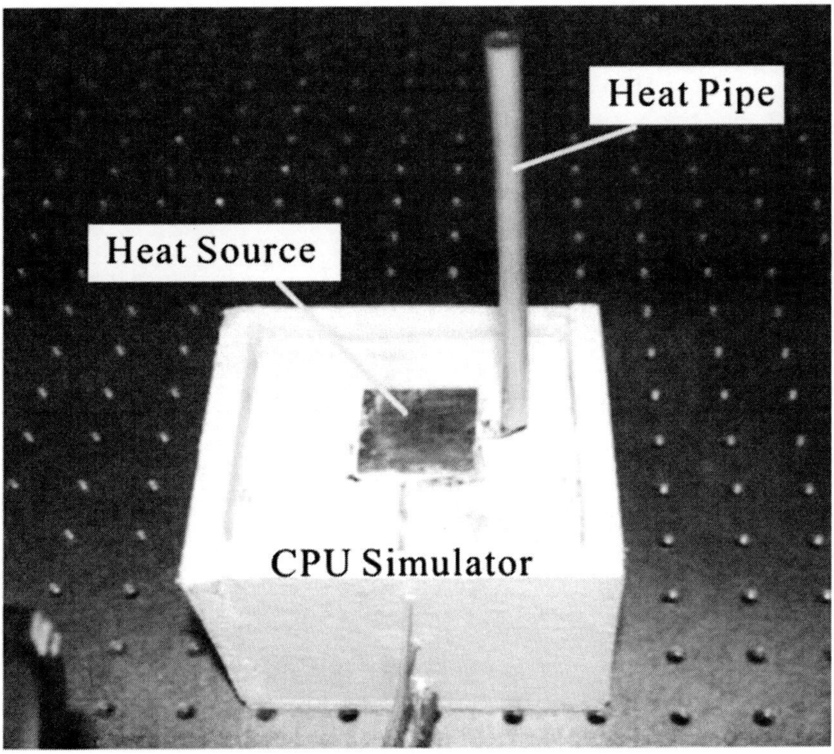

Figure 6. Photo of the heat pipe built on the CPU simulator.

In process, the copper pipe and the conventional heat pipe are individually connected with the CPU simulator for developing temperature measurements without outside cooling. These measurements can determine heat transfer properties of the copper pipe and the heat pipe. In principle, heat transfer property of the copper pipe is continuous and the heat pipe is discrete. After that, a commercial heat pipe cooler is connected with the CPU simulator for determination of the surface temperature distribution on the heat pipe. These measurements will show heat transfer property of the heat pipe influenced by the outside cooling fin. In comparison with the commercial heat pipe cooler uses the heat pipe to be immersed into water for visualization of the developing heat flux. The water is applied as the outside cooling medium in this work. The time-varying schlieren photos will show heat transfer property of the heat pipe.

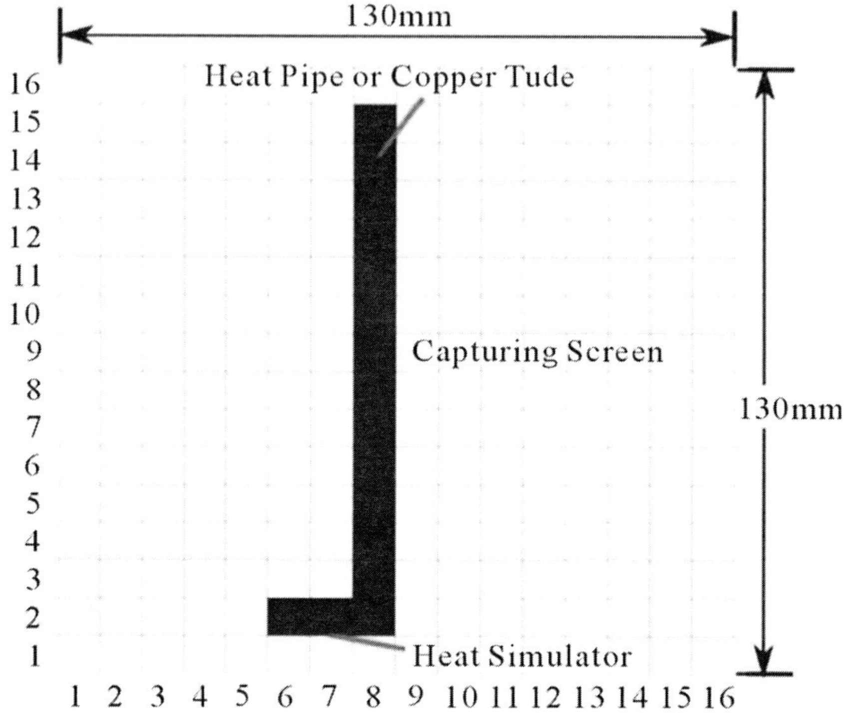

Figure 7. Capturing screen of the infrared thermal camera corresponding to the position of the heated heat pipe.

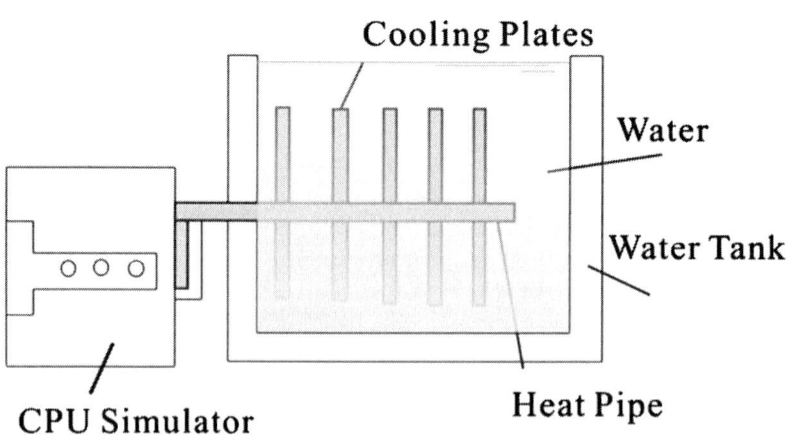

Figure 8. Schematic setup of the heat pipe in connection with the cooling plates.

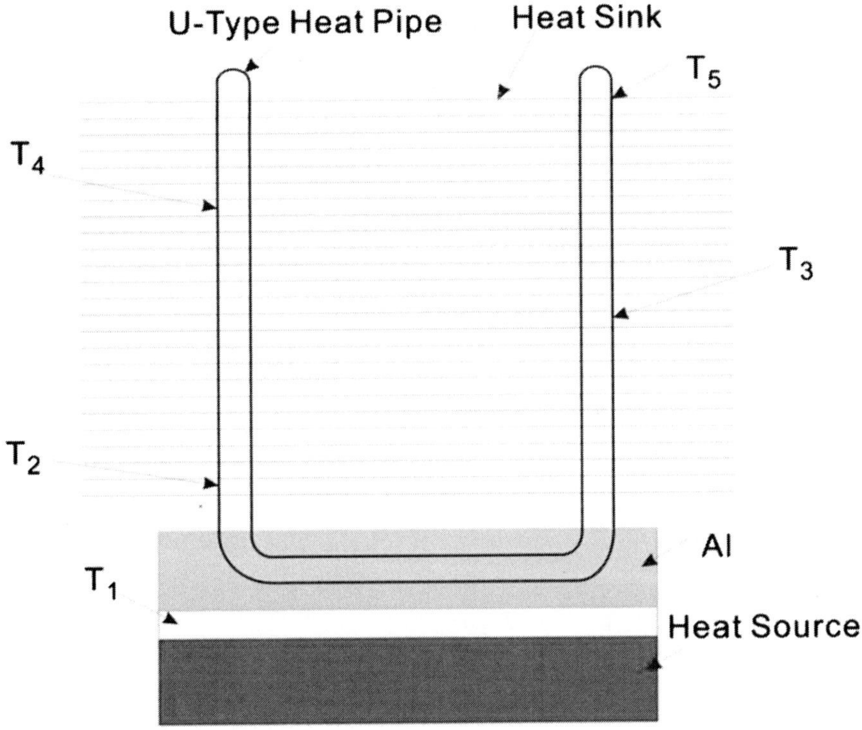

Figure 9. Schema of the measured temperature points in the heat pipe cooler.

4. RESULTS AND DISCUSSION

Experimental data for discussing discrete heat transfer behavior of the heat pipe in comparison with the copper pipe's continuous property are shown in Figures 11 and 12. Figure 11 shows the temperature distribution on the copper pipe heated under adiabatic condition by the 300W CPU simulator in 117sec [9]. The arabic positional numerals shown in Figures 11 and 12 are the measured positions corresponding to the distance from the heat source. The temperature change on the copper pipe is continuous. Figure 11 shows that the temperature distribution on the copper pipe gradually reduces away from the heat source. This result agrees to the traditional thermal conduction theorem and also satisfies the Fourier's thermal conduction equation. In comparison with the result of the heat pipe in

Figure 11, it is found that the change of surface temperature distribution on the heat pipe nearby the heat source is weaker than the endpoint which is under the same experimental condition. Figure 12 shows that the discrete temperature distribution on the heat pipe in middle field has the lowest temperature.

Figure 10. Photo of the measured heat pipe cooler.

The result shown in Figure 12 agrees to the theoretically described heat transfer behavior of the heat pipe. The evaporated working fluid saves latent heat from the heat source and expands directly to the remote condensed section. The heat transfer via the vapor expansion technism inside the heat pipe is relatively quick than via the thermal conduction through the outer copper shell. This special heat transfer technism let it have extremely large thermal conductivity. The temperature on positions 2 and 3 in Figure 12 appears unstability because vaporization was happening there.

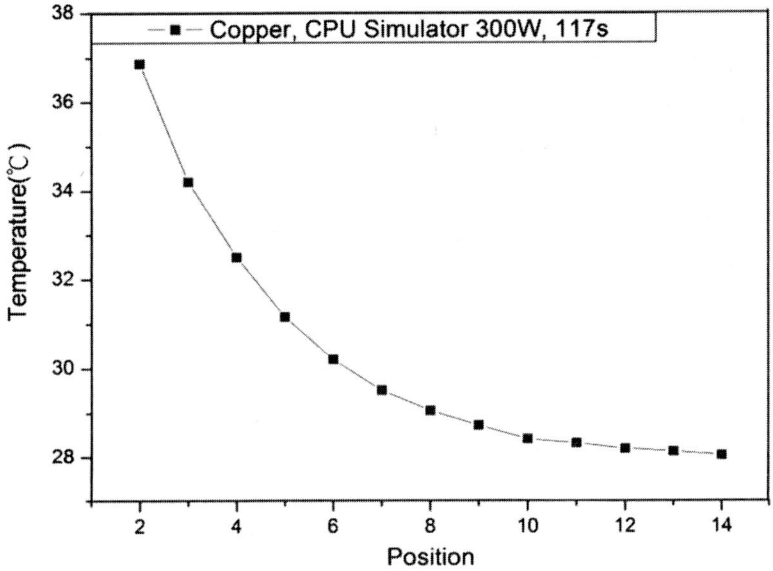

Figure 11. Relationship of temperature distribution on the bare copper pipe.

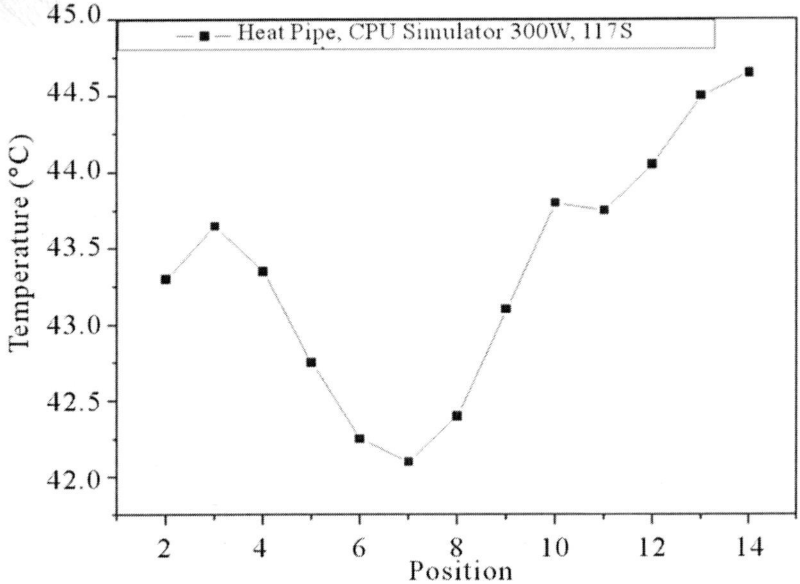

Figure 12. Relationship of temperature distribution on the bare heat pipe.

Figure 13 is the temperature distribution on the heat pipe which is built in the commercial heat pipe cooler. The heat pipe cooler has been heated by 30.8 W thermal power for 40 mins. The heat transfer behavior shown in Figure 13 is similar to the copper pipe in Figure 11. That means, the heat transfer behavior inside the heat pipe was altered by the addition of the cooling plates. In other words, the cooling plates changed the heat transfer behavior of the heat pipe.

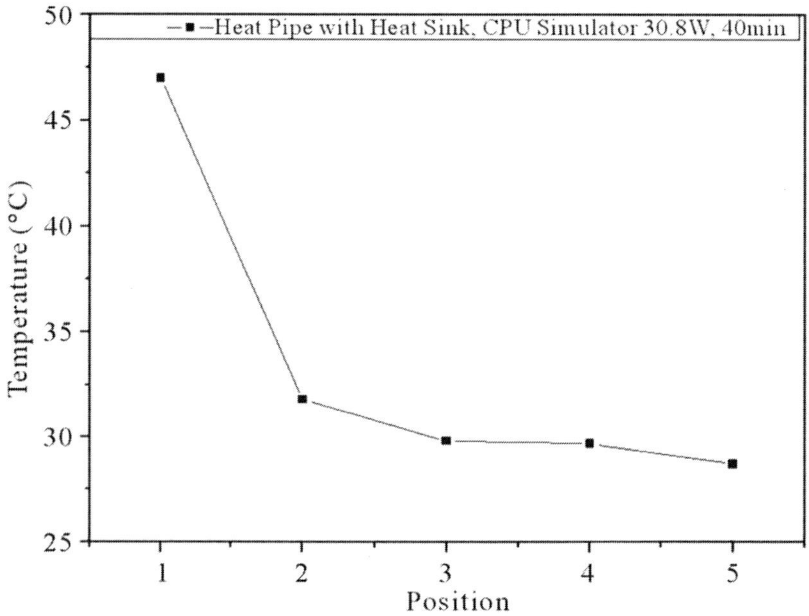

Figure 13. Relationship of temperature distribution on a commercial heat pipe cooler.

The special heat transfer technism of the heat pipe is further described by the direct observation of heat flux using the color schlieren technique. Water has larger thermal conduction coefficient $K_{water} \approx 2.526$ [W/mK] than air $K_{air} \approx 0.23$[W/mK] [11]. The water around the heat pipe is like the cooling apparatus while the heat pipe is immersed into the water tank for heat flux visualization [12]. The heat pipe transferred the heat into the water and changed the density distribution of the water, which was visualized by the color schlieren technique. Figures 11 exhibit four time sequent color schlieren images with the heated bare heat pipe. Figures 11 show that the horizontal dark bar in the middle is the heat pipe and the separation line upper the heat pipe is the water level. Figure 14(a) is at the

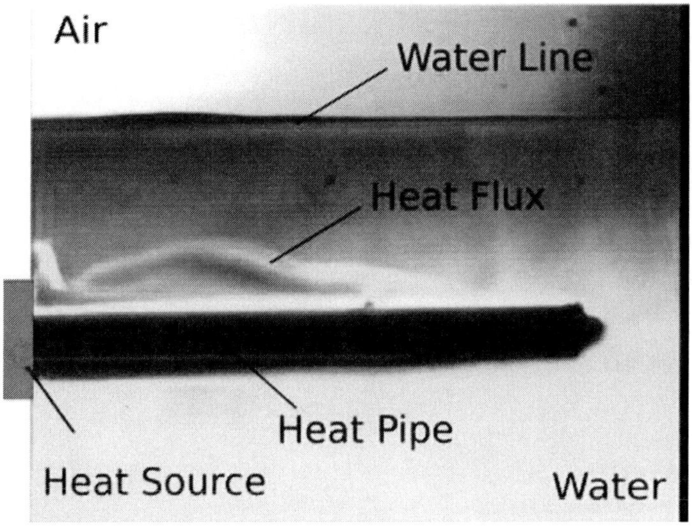
Air
Water Line
Heat Flux
Heat Pipe
Heat Source
Water

(a)

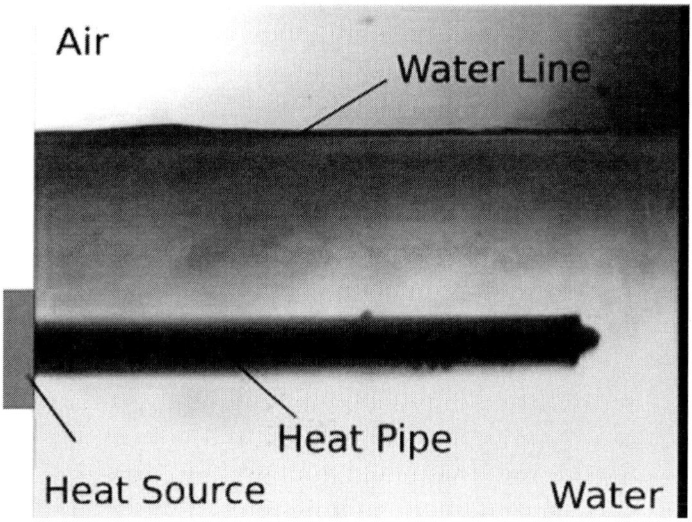
Air
Water Line
Heat Pipe
Heat Source
Water

(b)

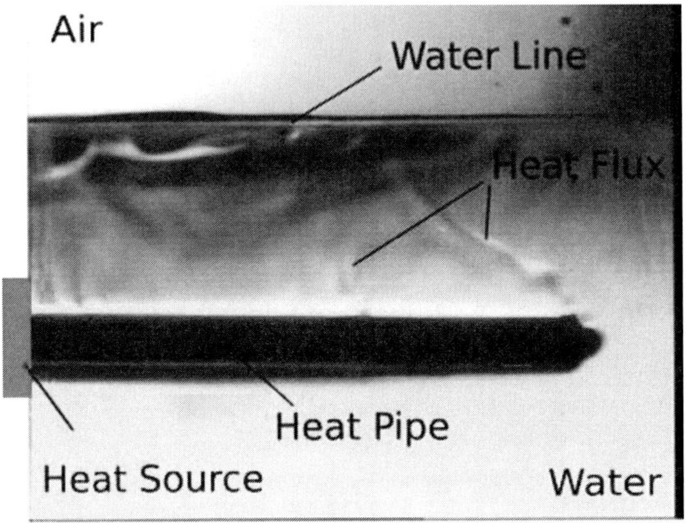

(c)

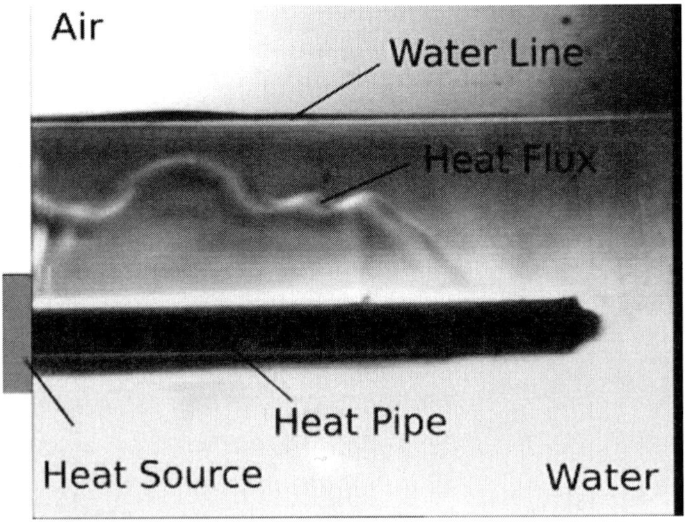

(d)

Figure 14. The heated time sequent color schlieren image with the heat pipe. (a) The heated time is 0sec. (b) The heated time is 15sec. (c) The heated time is 18sec. (d) The heated time is 22sec.

begin and shows no heat flux around the heat pipe. Figures 14(b) and (c) show clearly the development of the heat flux away from the heat source. Figures 14(a) and (b) exhibit indirectly that the heat transfer inside the heat pipe happens from the heat source and runs away continuously. Around the full heat pipe shows clearly the continuous heat flux from the left image to the right image. That means, the theoretically adiabatic middle section appears clearly thermal convection with the water. This significant result indicates further that the evaporated heat cannot be effectively brought to the remote condensed section, if the surroundings of the heat pipe has good thermal conductivity.

Heat pipe cooler uses normally aluminum plates as the cooling apparatus to take away heat from the heat pipe. The aluminum cooling plates have large thermal conduction coefficient $K_{Al} \approx 230[W/mK]$ [11]. In this work, the surrounding water plays the role of the integrated cooling plates in heat pipe cooler. Figure 11 indirectly indicates that the integration of cooling plates changes really the original heat transfer behavior of the bare heat pipe. This result completely agrees to the experimental data shown in Figure 10. The changed heat transfer behavior of heat pipe integrated with cooling plates reduces also its cooling efficiency. This paper indicates the fact that rebuilding the original heat transfer behavior of the bare heat pipe is the proper way for improving the cooling efficiency of the heat pipe cooler.

5. CONCLUSIONS

Temperature measurements using infrared thermography and thermocouple thermometer on the heat pipe as well as heat flux visualization around the heat pipe using color schlieren technique were successfully carried out. The results show that the heat transfer behavior of copper pipe is continuous, whereas heat pipe is discrete. The temperature distribution on heat pipe built-in heat pipe cooler shows good similarity to copper pipe. Time sequent color schlieren images show the same result as the temperature measurements. This work illustrates that the heat transfer behavior inside the heat pipe is changed to be similar to the copper pipe due to the integration with the cooling plates. Rebuilding the bare heat pipe's heat transfer behavior is the best solution for improving the cooling efficiency of the heat pipe cooler due to the original heat transfer behavior of the heat pipe has the best uniform temperature.

6. ACKNOWLEDGEMENTS

The authors would like to acknowledge the financial support from the National Science Foundation of Taiwan under Grant No. NSC97-2221-E-027-031.

REFERENCES

1. G. M. Grover, T. P. Cotter and G. F. Erickson, "Structures of Very High Thermal Conductance," Journal of Applied Physics, Vol. 35, No. 6, 1964, pp. 1990-1991.
2. F. J. Stenger, "Capillary Pumped Loop-CPL," NASA TM X-1310, 1996.
3. S. V. Vershini and Y. F. Maydanik, "Hysteresis Phenomena in Loop Heat Pipes," Applied Thermal Engineering, Vol. 27, No. 5-6, 2007, pp. 962-968.
4. A. Bejan and E. Sciubba, "The Optimal Spacing of Parallel Plates Cooled by Forced Convection," International Journal of Heat and Mass Transfer, Vol. 35, No. 12, 1992, pp. 3259-3264.
5. R. H. Yeh and M. Chang, "Optimum Longitudinal Convective Fin Arrays," International Communications in Heat and Mass Transfer, Vol. 22, No. 3, 1995, pp. 445-460.
6. P. L. Dhar and C. P. Arora, "Optimum Design of Finned Surface," Journal of the Franklin Institute, Vol. 301, No. 4, 1976, pp. 379-392.
7. B. Kundu and P. K. Das, "Performance and Optimum Design Analysis of Convective Fin Arrays Attached to Flat and Curved Primary Surfaces," International Journal of Refrigeration, Vol. 32, No. 3, 2009, pp. 430-443.
8. H. S. Kou, J. J. Lee and C. Y. Lai, "Thermal Analysis and Optimum Fin Length of a Heat Sink," Heat Transfer Engineering, Vol. 24, No. 2, 2003, pp. 18-29.
9. C. C. Ting, J. N. Lee and C. C. Chen, "Heat Transfer Characterizations of Heat Pipe in Comparison with Copper Pipe," Journal of Heat Transfer, Vol. 131, No. 3, 2009, pp. 1-6.
10. C. C. Ting and C. C. Chen, "Developing the Coaxial Dual-Pipe Heat Pipe for Applications on Heat Pipe Cooler," Journal of Heat Transfer, Vol. 133, No. 9, 2011.

11. J. M. Coulson and J. F. Richardson, "Fluid Flow, Heat Transfer and Mass Transfer," Butterworth-Heinemann Ltd., Oxford, 1999.

12. H. Kleine, "Application and Limitation of a Schlieren System," RWTH-Aachen Germany, Report, No. WLS 12823, 1994.

8

CHAPTER

EFFECTS OF INHOMOGENEITIES ON HEAT AND MASS TRANSPORT PHENOMENA IN THERMAL BRIDGES

Paola Iodice [1], Nicola Massarotti [2] and Alessandro Mauro [2,*]

[1] *Università Telematica Pegaso, Centro Direzionale, Isola A3, 80143 Napoli, Italy*

[2] *Dipartimento di Ingegneria, Università degli Studi di Napoli "Parthenope", Centro Direzionale, Isola C4, 80143 Napoli, Italy*

ABSTRACT

The interest of calculating the effects of thermal bridges in buildings energy consumption is growing, due to recent energy saving regulations applied in different countries. The widespread use of insulating materials to reduce energy requirements of buildings, often employed for intermediate insulation of the building envelope, makes thermal bridges a crucial point in the energy analysis of building envelopes. Furthermore, heat losses through thermal bridges often lead to building pathologies due to moisture condensation. Therefore, thermal bridges need to be correctly characterized in the building design stage in order to reduce heat losses and avoid materials degradation. The authors numerically simulate, by using finite elements, the steady-state and dynamic three-dimensional (3D) heat and vapor transport in inhomogeneous thermal bridges and building envelopes. The aim of the present work is to show the importance of taking into account the presence of inhomogeneities (i.e., metal stud) in building materials for the calculation of actual heat losses and water condensation in 3D thermal bridges. The obtained heat transfer results are verified against the reference data of the technical standard UNI EN ISO 10211. The proposed microscopic approach is essential to calculate the actual heat losses of three-dimensional thermal bridges and building envelopes and to overcome condensation problems.

Keywords: inhomogeneity; dynamic; conduction; diffusion; thermal bridge; thermal dispersions

1. INTRODUCTION

The concept of energy saving includes a variety of techniques aimed at reducing the energy consumption needed for the different human activities. Nowadays, a great deal of attention has been given to efficient energy use in the heating and cooling sector, and building envelopes play a crucial role in the reduction of the energy consumption and the related costs. In fact, building heating and cooling are responsible for about 40% of the final energy consumption in the European Union (EU) and, as a consequence, for a considerable amount of total CO_2 emissions [1]. The attention to the issue of energy efficiency in buildings is demonstrated by international and national directives [2,3,4,5], which describe the procedure for energy certification of buildings, and require member states to ensure that new and existing buildings meet minimum energy performance requirements, monitoring the actual energy demand or the estimated energy necessary to meet the various needs related to standard building uses, such as heating and cooling.

In this perspective, the methodology used to calculate the heat losses through the building envelope plays a crucial role, as well as the correct determination of the constitutive properties of the materials used. In many cases, one-dimensional and two-dimensional models do not correctly reproduce the heat transfer and vapor diffusion through the components of the building enclosure, but it is necessary to employ three-dimensional numerical models capable of solving the partial differential equations through the actual building envelope, characterized by the presence of geometric and structural non-uniformities, defined as thermal bridges.

These, characterized by a multi-dimensional heat flow and a consequent lowering of internal surface temperatures, are also those structures in which condensation and mold growth phenomena occur, due to incorrect design of the building envelope from the hygrothermal point of view. In this regard, the verification of hygrothermal performance [6] has the purpose of determining the possible presence of water condensed in a part of the examined structure, and then proceed to its correct design in order to avoid such critical phenomenon. Despite the requirements concerning the insulation of buildings, specified by national and international regulations, thermal bridges are a weak spot in constructions, since are often not dealt with properly. For example, the presence of inhomogeneities (i.e., metal stud) inside construction materials is often not taken into account, despite the consequent results could be very different, as shown in this work. These aspects lead to underestimate the heat losses during the design stage, therefore providing lower insulation levels for the building envelope, with consequent higher actual energy consumption, if compared to the estimates [7]. It should be pointed out that the use of different calculation methods of thermal bridges, provided in standards [8,9], can lead to different energy performance classes of a building and to percentage gaps on the energy requirements in the heating season, calculated to be more than 20% [10].

In this scenario, the use of numerical tools, required by the standards for the characterization of thermal bridges [9], should be extended to the verification of hygrothermal performance of these structures, given the complexity of solving the governing equations analytically. In these analyses, the thermal and hygrometric properties employed for each material of the building envelope assume a crucial role. These properties are available for most building components [11,12,13], or must be certified by the manufacturer and are commonly used in design calculations. However, problems may arise for inhomogeneous materials, such as concrete, heavily used in the existing building envelopes. In fact, the presence of steel stud strongly influences the thermal conductivity and the hygrometric properties of the inhomogeneous concrete structure. The adoption of the value of thermal conductivity of simple concrete may cause significant errors in the calculation of the actual heat losses. Furthermore, neglecting the presence of the metal stud, some areas of possible water condensation may be ignored, thus preventing the subsequent corrective action.

Therefore, the authors' aim in this work is to show the importance of taking into account the presence of inhomogeneities inside construction materials. This approach, referred to as "microscopic", requires to reproduce the inhomogeneities (i.e., metal stud), with their physical properties, inside the geometry under investigation.

The present paper is structured as follows: the next section provides a review of the available literature on numerical modelling of thermal bridges, and introduces the novelties of the present paper, then the mathematical model employed is described in section three. The fourth section shows the influence of the inhomogeneities inside building structures, while the fifth section reports the verification of the obtained results for a benchmark problem. The sixth section presents the main results obtained for the three-dimensional thermal bridge taken into account and for a room envelope, while some conclusions are drawn in the last section.

2. REVIEW OF THE AVAILABLE LITERATURE

The importance of quantifying the heat loss through thermal bridges is well recognized in engineering practice, and technical standards provide a way to approximate the behavior of thermal bridges under steady design conditions. However, in order to take into account actual transient conditions, a number of works have studied thermal behavior of linear thermal bridges, to analyze their contribution to the total energy performance of buildings. The phenomenon depends on several factors, such as external and internal conditions, level of insulation, geometry of the thermal bridge and materials used, structure and use of building, and the results of calculations, unlike reality, often depend on the model used to calculate its effects on building energy demand [14]. Therefore, estimates of the actual weight of thermal bridges on the energy demand of buildings can vary from 5%, in the case of retrofitting the exterior of the building envelope, to 39% in well insulated single family houses without proper thermal bridge treatment [14].

Due to the complexity of three-dimensional (3D) unsteady simulations, most studies use one-dimensional (1D) and two-dimensional (2D) approaches or steady-state regime. A review of methods for dynamic modelling of thermal bridges in building envelopes is reported in [15]. Moreover, most commercial software for the dynamic simulation of thermal performance of buildings assume 1D heat transfer neglecting 2D dynamic heat conduction through the thermal bridges of building enclosures [16].

An earlier work of Kosny and Christian [17] showed that thermal resistance of a concrete wall reinforced with metal stud may decrease of about 48% because of the steel-concrete thermal bridge effects. The authors employed a finite difference algorithm for the simulation of transient heat transfer in building walls and wall corners. Moreover, Kosny and Kossecka [18] demonstrated, by using a 3D finite difference algorithm, that the thermal resistance of walls, calculated neglecting the thermal bridge effects and adopting the one-dimensional approach, might be overestimated of more than 44%. Gao et al. [16] used a reduction technique to simulate three-dimensional heat transfer. The work focused on the study of the dynamic behavior of thermal bridges in a simplified form by using a commercial simulation software. The study showed that thermal bridges are responsible for an increase in the overall heat loss of the building taken into account from 9% to 19%. Mao and Johannesson [19] developed a code based on a frequency response approximation method to calculate temperatures and heat fluxes in thermal bridges, by using the finite element discretization technique. Déqué et al. [20] integrated a 2D dynamic modeling of thermal bridges into a commercial software, showing that this approach can lead to a difference of 7% in heat losses evaluated with the use of tabulated values. An interesting 3D numerical study on the influence of outdoor environmental conditions on building energy performance has been carried out by He et al. [21] by using a proprietary algorithm, but neglecting the presence of thermal bridges in the building envelope. Recently, Tadeu et al. [1] proposed a 2D dynamic model, based on the boundary element method in the frequency domain, to characterize wall corners and estimate their influence on the thermal performance of building components. Theodosiou et al. [22] carried out 3D heat transfer simulations of cladding systems for building facades, showing that neglecting the point thermal bridge effect in these kind of systems can lead to a significant underestimation (from 5% to 20%) of actual heat flows.

Some combined experimental and numerical works are also available in the literature. Nussbaumer et al. [23] analyzed both experimentally and numerically a concrete wall with external insulation of polystyrene boards, containing vacuum insulation panels, demonstrating the thermal improvement obtainable with this type of insulation. Wakili et al. [24] reproduced numerically, with a 3D commercial code, steady-state temperature field in a balcony board with integrated glass fiber reinforced plastic elements. Numerical results, combined with experimental data, proved that a reduction of heat losses are obtainable by using glass fiber reinforced elements instead of conventional steel rods. Zalewski et al. [25] employed a 3D commercial code for steady state analysis of thermal efficiency of complex building walls, focusing on the estimation of heat losses from thermal bridges. The authors also carried out an experimental analysis by using infrared thermography and heat flow meters for quantitative characterization of the losses through the envelope. A recent work by Lorenzati et al. [26] investigated the influence of thermal bridges on the overall energy performance of Vacuum Insulation Panels (VIP). They showed that the use of equivalent values for thermal conductivity of VIP could lead to significant errors in the estimation of thermal bridges' effects.

The above literature review proves the technical-scientific interest for the calculation of heat losses through thermal bridges, given the significant influence that these components of a building have on the global heat losses of buildings. To the authors' knowledge, some aspects on this subject are not dealt with in depth in the available literature. In particular, the above literature review shows that a microscopic approach that takes into account the presence of inhomogeneities inside construction materials has not been employed coupled with 3D numerical simulations of thermal bridges under dynamic conditions. Furthermore, the verification of hygrothermal performance in presence of inhomogeneous materials, which is a crucial aspect for building structures, has not been extended to 3D thermal bridges. On the basis of these considerations, in this work the authors, experienced in the field of numerical simulation of complex thermo-fluid-dynamics problems and energy saving systems [27,28,29,30], investigate the influence of inhomogeneities inside construction materials simulating transient heat conduction and vapor diffusion in 3D thermal bridges.

3. MATHEMATICAL MODEL

The mathematical model employed in the present work for the simulation of heat and mass transfer through a three-dimensional thermal bridge consists in the unsteady conservation equations for mass and energy in a homogeneous permeable medium. The governing equations have been solved by using the software Comsol Multiphysics. To employ the present microscopic approach, the governing equations are therefore written for each isotropic homogeneous material that is found in a typical building envelope:

Heat transfer by conduction

$$\rho c \frac{\partial T}{\partial \vartheta} = \lambda \nabla^2 T \tag{1}$$

Vapor diffusion

$$\frac{\partial p_v}{\partial \vartheta} = \delta \nabla^2 p_v \tag{2}$$

Therefore, in order to take into account the presence of metal stud in the thermal bridge, Equation (1) is also solved in the domain occupied by the metal.

The set of partial differential Equations (1) and (2) are solved with the following boundary and initial conditions for energy and mass:

$$T(x,y,z,\vartheta=0)=T_0;\quad p_v(x,y,z,\vartheta=0)=p_{v,0}\quad \text{in the whole computational domain}$$

$$h_{c,i}\left[T_{\infty,i}-T_{P,i}(x,y,z,\vartheta)\right]=-\lambda\left.\frac{\partial T(x,y,z,\vartheta)}{\partial n}\right|_{P,i};\quad p_v=p_{v,i}\quad \text{on the interior surfaces}$$

$$h_{c,e}\left[T_{\infty,e}-T_{P,e}(x,y,z,\vartheta)\right]=-\lambda\left.\frac{\partial T(x,y,z,\vartheta)}{\partial n}\right|_{P,e};\quad p_v=p_{v,e}\quad \text{on the exterior surfaces}$$

$$\left.\frac{\partial T(x,y,z,\vartheta)}{\partial n}\right|_{P}=0;\quad \left.\frac{\partial p_v(x,y,z,\vartheta)}{\partial n}\right|_{P}=0\quad \text{on the lateral surfaces} \tag{3}$$

where $T0$ and $pv,0$ are the initial temperature and partial pressure of vapor, respectively, hc,i and hc,e are the convective heat transfer coefficients of the interior and exterior environment, respectively, $T\infty,i$ and $T\infty,e$ are the interior and exterior environment temperatures, respectively. TP,i and TP,e are the temperatures of the interior and exterior surfaces, respectively, pv,i and pv,e are the partial pressures of vapor in the interior and exterior environment. The mathematical model described above is solved by using the well known Finite Element Method [31,32].

The procedure for the verification of the hygrometric performance (Glaser's method), based on the technical standard [6], is here applied to thermal bridges. In order to avoid interstitial and superficial water condensation, the following condition must be satisfied in each point of the structure:

$$p_v < p_{sat}(T) \tag{4}$$

where $psat(T)$ is the saturation pressure of water at temperature T.

The relation between temperature and saturation pressure of water is given in [6] as:

$$p_{sat}(T)=610.5\,e^{\left(\frac{17.269\cdot T}{237.3+T}\right)}\quad \text{for } T\geqslant 0\,^{\circ}\mathrm{C}$$

$$p_{sat}(T)=610.5\,e^{\left(\frac{21.875\cdot T}{265.5+T}\right)}\quad \text{for } T<0\,^{\circ}\mathrm{C} \tag{5}$$

in which the temperature must be expressed in degrees Celsius (°C).

In common practice, the conditions characterized by a vapor partial pressure, pv, larger than saturation pressure of water at a prescribed temperature, $psat(T)$, are not physical, because saturation pressure is the maximum vapor pressure compatible with the prescribed temperature value, T. Therefore, inside the computational domain, wherever $[pv-psat(T)]\geq 0$, condensation occurs. The Glaser's method applied to one-dimensional problems, such as those assumed in most building walls, allows to calculate the amount of water condensed in the wall as the difference between diffusive vapor flux entering and leaving the different layers of the wall [33,34]. This same type of analysis can be extended to three-dimensional problems, where condensation occurs in a volume delimited by surfaces (defined as A and B in the followings) on which the condition $[pv-psat(T)]|A,B=0$ is verified, while inside the volume the condition $[pv-psat(T)]\geq 0$ occurs. Based on these considerations, the present numerical procedure is developed in order to estimate the rate of condensing water, $m^{\cdot}w$, in a three-dimensional structure when condition Equation (4) is not verified, from the difference between the calculation of vapor flux and vapor flux at saturation conditions, as follows:

$$\dot{m}_w = g_v|_A - g_{sat}|_B \tag{6}$$

where $gv|A$ is the diffusive vapor flux entering the condensation volume through surface A, while $gsat|B$ is the diffusive vapor flux calculated at saturation conditions leaving the condensation volume through surface B:

$$g_v|_A = -\delta \int_A \nabla p_v \cdot n\, dA$$

$$g_{sat}|_B = -\delta \int_B \nabla p_{sat}(T) \cdot n\, dB \tag{7}$$

where surfaces A and B cannot be defined a-priori, since they depend on the temperature distribution in the considered geometry.

4. INFLUENCE OF INHOMOGENEITIES INSIDE BUILDING STRUCTURES

Materials employed in building applications are commonly considered to be homogeneous and isotropic, and their thermal properties are often dependent on temperature and density. However, when inhomogeneous and anisotropic media are taken into account, the physical properties are different because of the inhomogeneities present inside the material.

In this section, the authors take into account two typical constructive elements: a concrete pillar with metal stud and a brick-concrete slab, in order to analyze the influence of the inhomogeneities on the heat flux transferred by these structures. The first case is presented in Figure 1, which shows the three-dimensional profile of a concrete pillar (0.3×0.5×3 m3) with and without the metal stud, together with section and stud details. Two different stirrup spacing are taken into account, 0.2 m and

0.1 m, in order to show the dependence of the heat flux on the construction characteristics. The materials considered are: (i) concrete of natural aggregates with closed structure and with a thermal conductivity λco=1.2 W/(m·K); (ii) steel, used as metal stud, with a thermal conductivity λs=50 W/(m·K) [11,12,13]. In order to calculate the heat flux transferred by different pillar configurations, it is necessary to reproduce the problem numerically under steady-state and specified boundary conditions. The boundary conditions employed are:

$$-\lambda\frac{\partial T}{\partial n} = h_{c,i}\,(T_{\infty,i} - T_{P,i}) \quad \text{with} \quad h_{c,i} = 7.7\,\text{W}/(\text{m}^2\cdot\text{K});\; T_{\infty,i} = 20\,^{\circ}\text{C} \quad \text{on the interior surface}$$

$$-\lambda\frac{\partial T}{\partial n} = h_{c,e}\,(T_{\infty,e} - T_{P,e}) \quad \text{with} \quad h_{c,e} = 25\,\text{W}/(\text{m}^2\cdot\text{K});\; T_{\infty,e} = 2\,^{\circ}\text{C} \quad \text{on the exterior surface}$$

$$\frac{\partial T}{\partial n} = 0 \quad \text{on the lateral surfaces} \tag{8}$$

where the convective heat transfer coefficients are taken from technical standard [35], while the interior temperature is provided in [36,37]. The value of the exterior temperature considered is the design value referred to the city of Napoli in Italy, given in the technical standard [38].

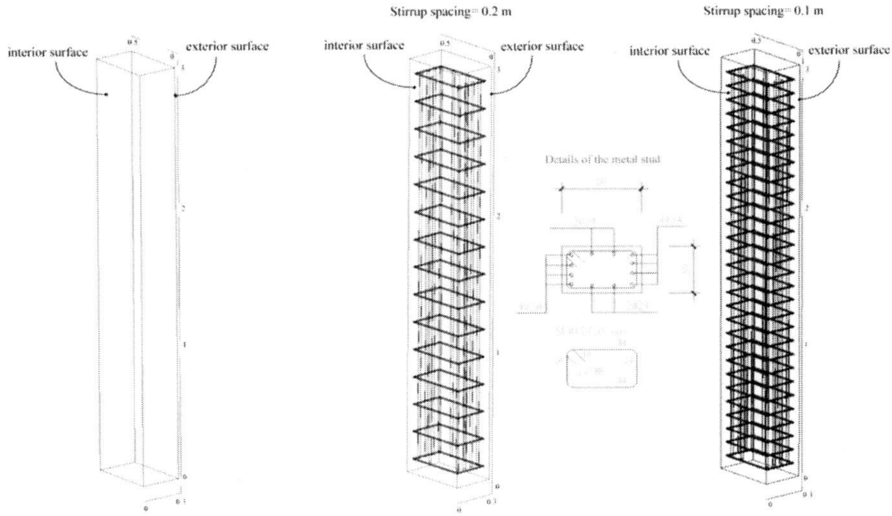

Figure 1. Three-dimensional profile of a concrete pillar (0.3×0.5×3 m3) without metal stud (**left**), with metal stud—stirrup spacing of 0.2 m (**center**), and with metal stud—stirrup spacing of 0.1 m (**right**).

The simulation of steady-state heat transfer allows to calculate, in the post-processing stage, the heat flux transferred from the exterior environment to the interior one. Table 1 reports the heat flux calculated for the three concrete pillar configurations. The presence of the metal stud in the concrete pillar leads to an increase of 6.8% of the heat flux if the metal stud with stirrup spacing of 0.2 m is considered, and 8.9% if the metal stud with stirrup spacing of 0.1 m is considered. From the analysis of the data presented in Table 1, it is evident that the presence of a steel stud in the concrete pillar has an influence on the heat flux transferred from the structure. This calculation should be carried out each time the amount of steel in the pillar changes.

Table 1. Heat flux transferred from the different configurations of pillar and slab.

	Heat flux (W/m^2)	
Concrete pillar	42.9	
Concrete pillar with metal stud—stirrup spacing 0.1 m	45.8	
Percentage difference		6.8%
Concrete pillar with metal stud—stirrup spacing 0.2 m	46.7	
Percentage difference		8.9%
Concrete slab	47.6	
Brick-concrete slab	33.3	
Percentage difference		−30%

The second case, presented in Figure 2, shows the three-dimensional profile of a concrete slab (0.25×1×1.8 m3) and of a brick-concrete slab, together with the details of the brick-concrete section. The materials considered are (i) concrete of natural aggregates with closed structure with a thermal conductivity λco=1.2 W/(m·K); (ii) steel rods with a thermal conductivity λs=50 W/(m·K); (iii) bricks with a thermal conductivity λb=0.4 W/(m·K) [11,12,13].

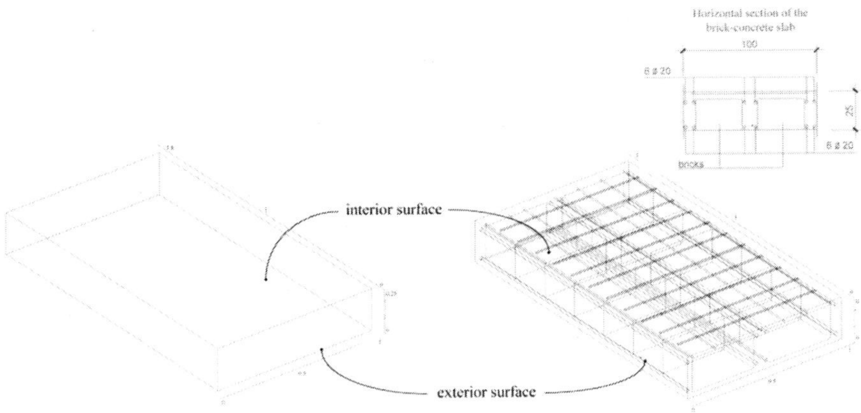

Figure 2. Three-dimensional profile of a concrete slab (0.25×1×1.8 m3) (**left**) and of a brick-concrete slab (**right**).

The boundary conditions employed are represented by Equations (8). The values of heat flux calculated by means of numerical simulation are reported in Table 1. It should be pointed out that the presence of the bricks strongly influences the heat flux transferred from the interior to exterior surface of the slab. In fact, the value of this parameter decreases of 30% with respect to the case of pure concrete (Table 1).

These results highlight the difference in terms of heat flux due to the presence of inhomogeneities inside construction materials, by using a microscopic approach. However, in order to correctly characterize a building structure in terms of both heat transfer and vapor diffusion, unsteady simulations should be carried out. In addition, the presence of inhomogeneities must be necessarily taken into account, by using the present microscopic approach, as shown for the two configurations of thermal bridges and for the room envelope presented in Section 6. With this approach, the actual building structure is numerically reproduced, and the physical properties of the materials reported in the technical standards [11,12,13] are considered. Therefore, the use of non-physical parameters, such as equivalent values of physical properties, is avoided. Before showing these results, the required verification of the numerical model is presented in the next section.

5. VERIFICATION OF THE NUMERICAL MODEL

The results obtained by using the present numerical model are verified against the reference data available in technical standard [9] for steady-state heat transfer through a three-dimensional thermal bridge. The geometry of the benchmark problem considered and the description of the surfaces with different boundary conditions are depicted in Figure 3. The computational domain consists of a three-dimensional structure that separates three environments kept at different temperatures. More details necessary for defining the geometry are available in reference [9]. The boundary conditions employed in the simulations, with reference to the surfaces highlighted in Figure 3, are:

$$
\begin{aligned}
&-\lambda \frac{\partial T}{\partial n} = h_{c,i1}\,(T_{\infty,i1} - T) \quad \text{with} \quad h_{c,i1} = 5\,\text{W}/(\text{m}^2 \cdot \text{K})\,;\; T_{\infty,i1} = 15\,^{\circ}\text{C} \quad \text{on the interior (green) surfaces} \\
&-\lambda \frac{\partial T}{\partial n} = h_{c,i2}\,(T_{\infty,i2} - T) \quad \text{with} \quad h_{c,i2} = 5\,\text{W}/(\text{m}^2 \cdot \text{K})\,;\; T_{\infty,i2} = 20\,^{\circ}\text{C} \quad \text{on the interior (blue) surfaces} \\
&-\lambda \frac{\partial T}{\partial n} = h_{c,e}\,(T_{\infty,e} - T) \quad \text{with} \quad h_{c,e} = 20\,\text{W}/(\text{m}^2 \cdot \text{K})\,;\; T_{\infty,e} = 0\,^{\circ}\text{C} \quad \text{on the exterior (pink) surfaces} \\
&\frac{\partial T}{\partial n} = 0 \quad \text{on the lateral (grey) surfaces}
\end{aligned}
\tag{9}
$$

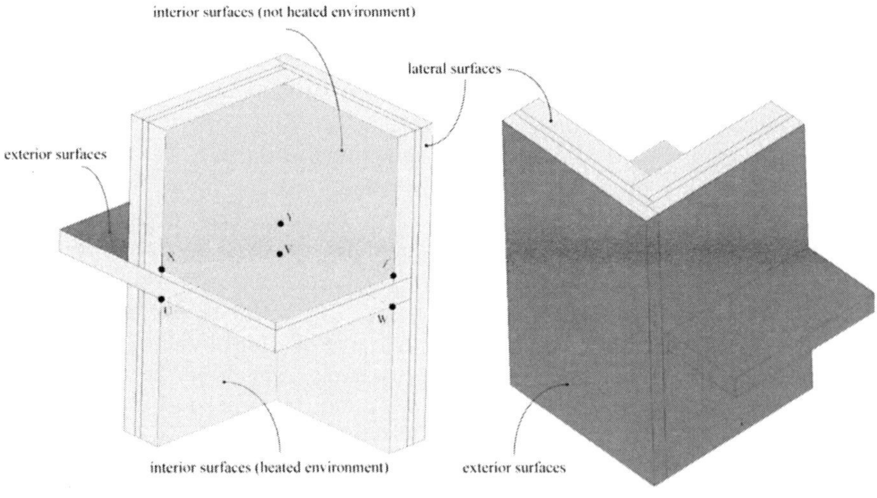

Figure 3. Computational domain considered: front view (**left**) and rear view (**right**) of the 3D thermal bridge.

The computational grid employed is made of 72010 nodes and 383715 tetrahedral elements. The mesh has been chosen after grid sensitivity analysis, in order to ensure that all the calculated values are mesh independent. Grid independency is considered to be achieved when the maximum difference between the quantities of interest calculated with a mesh and those obtained with the next finer grid is smaller than 1%.

The verification of the results against reference data [9] is presented in Table 2. The points and surfaces considered for temperature calculation are reported in Figure 3. In order to consider the numerical model verified, the technical standard [9] requires a difference between the calculated temperatures and reference data to be less or equal to 0.1 K, and a difference on the obtained heat flow rate and reference values less equal than 2%. From the analysis of Table 2, it is evident that the present results are in perfect agreement with the data provided in [9], and the above requirements are verified.

Table 2. Verification of the results against reference data [9].

Temperature (°C)			
Points	**Present Solution**	**Reference Data [9]**	**Difference**
U	12.90	12.90	0.00
V	11.31	11.30	0.01
W	16.41	16.40	0.01
X	12.56	12.60	0.04
Y	11.08	11.10	0.02
Z	15.28	15.30	0.02
Heat Flow Rate (W)			
Surfaces	**Present Solution**	**Reference Data [9]**	**Difference (%)**
interior blue surfaces	45.49	46.30	1.75
interior green surfaces	13.81	14.00	1.36
exterior pink surfaces	59.48	60.30	1.36

6. NUMERICAL RESULTS

In the present work, the authors have simulated both steady-state and unsteady conduction heat transfer and diffusion of vapor in a three dimensional (3D) thermal bridge. The geometry considered for the numerical analysis is similar to the one proposed in the technical standard [9], opportunely modified to reproduce a structure adopted in actual constructions. In particular, a pillar (0.3×0.3×2.3 m3) is inserted at the corner, and the presence of interior and exterior plasters and of tiles is considered.

Furthermore, in order to show the importance of taking into account the presence of internal inhomogeneities of the materials employed in constructions to correctly estimate heat losses and condensation, two different cases are considered: the pillar and the slab are made of concrete of natural aggregates with closed structure (case 1); a steel stud in the pillar and a brick-concrete slab are considered (case 2). The 3D structures for the two cases of modified thermal bridge are reported in Figure 4 and Figure 5, respectively, while the wall and slab sections are reported in Figure 6. The physical properties of the materials employed in the modified thermal bridge, taken from the technical standards [11,12,13], together with their geometrical characteristics, are reported in Table 3. The geometry of thermal bridge corresponding to case 2 is the same as that studied by one of the authors in a recent paper [39], where the aim was to analyze experimentally and numerically the effects of innovative materials employed in building structures.

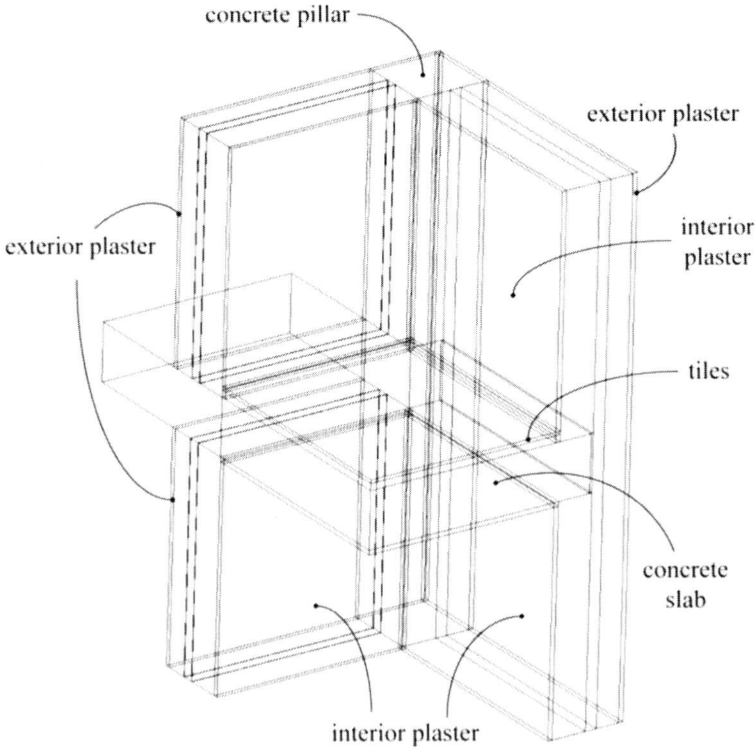

Figure 4. Three-dimensional profile of the modified thermal bridge with concrete pillar and slab (case 1).

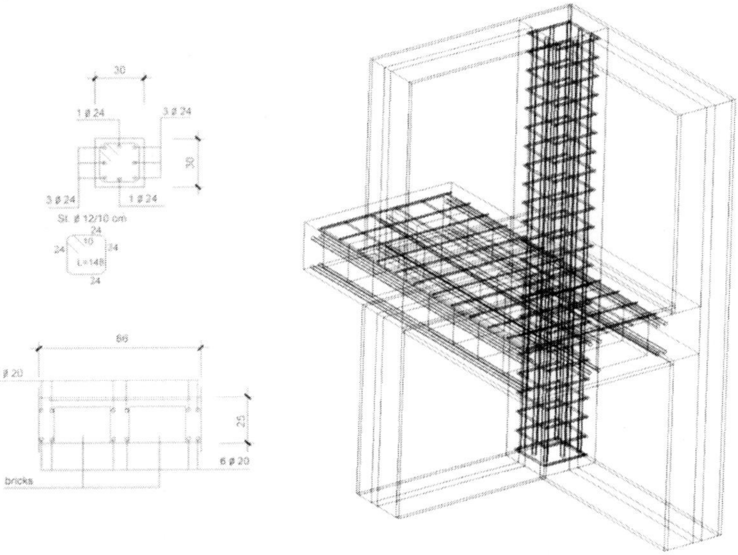

Figure 5. Three-dimensional profile of the modified thermal bridge with steel-concrete pillar and brick-concrete slab (case 2), with pillar and slab sections (**left**).

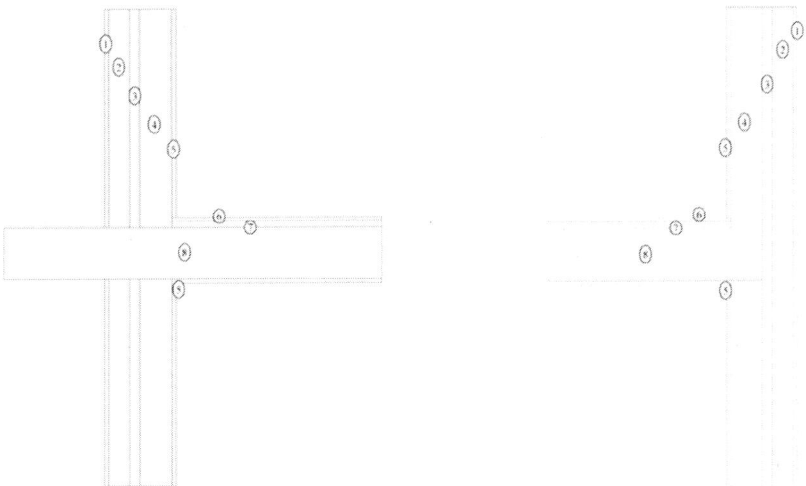

Figure 6. Longitudinal and transverse sections of the modified thermal bridge.

Table 3. Properties of the materials employed in the modified thermal bridge.

Material	Reference	Thermal Conductivity λ (W/(m · K))	Vapor Permeability $\delta \cdot 10^{12}$ (kg/(m · s · Pa))	Densiy ρ (kg/m^3)	Specific Heat Capacity c (J/(kg · K))	Thickness s (m)
Exterior plaster	n.1—Figure 6	0.70	18	1400	1000	0.02
Solid brick	n.2—Figure 6	0.72	27	1800	840	0.10
Insulation	n.3—Figure 6	0.04	3.2	30	1250	0.04
Hollow brick	n.4—Figure 6	0.30	27	800	840	0.12
Interior plaster	n.5—Figure 6	0.35	18	1200	1000	0.02
Malta	n.7—Figure 6	0.90	9.0	1800	1000	0.03
Tiles	n.6—Figure 6	1.4	9.0	2300	840	0.015
Concrete-pillar	Figure 4	1.2	3.0	2000	1000	0.30
Concrete-slab	n.8—Figure 6	1.2	3.0	2000	1000	0.25
Steel-stud	Figure 5	52	0.0	7800	500	Figure 5

In addition to the 3D thermal bridge, the authors analyze numerically a room of an apartment located at the top floor of a building.

6.1. Thermal Bridge under Steady-State Conditions

The boundary conditions employed for the solution of steady-state heat transfer are represented by Equations (9), but with different values of *hc,i1, hc,i2, hc,e*: 7.7 W/(m2·K), 7.7 W/(m2·K) and 25 W/(m2·K), respectively, as required in technical standard [35]. The boundary conditions employed for the solution of steady-state vapor diffusion, referred to the same surfaces considered for heat transfer and highlighted in Figure 3, are:

$$p_{v,i1} = 852\,\mathrm{Pa}\ (\text{corresponding to } \phi_{i1} = 50\%) \quad \text{on the interior (green) surfaces}$$

$$p_{v,i2} = 1168\,\mathrm{Pa}\ (\text{corresponding to } \phi_{i2} = 50\%) \quad \text{on the interior (blue) surfaces}$$

$$p_{v,e} = 366\,\mathrm{Pa}\ (\text{corresponding to } \phi_{e} = 60\%) \quad \text{on the exterior (pink) surfaces}$$

$$\frac{\partial p_v}{\partial n} = 0 \quad \text{on the lateral (grey) surfaces} \tag{10}$$

From the obtained numerical results, reported in Table 4 in terms of the heat flux transferred by the interior and exterior surfaces of the thermal bridge, it is possible to appreciate significant differences between case 1 and case 2. The difference on the heat flux calculated for case 1 and case 2, due to the presence of inhomogeneities in case 2, reaches the 16%. Obviously, these results depend on the geometry and on the construction characteristics of the structures considered in the simulations, and the percentage gap can increase for larger amount of steel or smaller bricks. Table 4 also reports the results obtained when the properties corresponding to reinforced concrete are used (case 1.b), taken from the technical standard [13]. In particular, a thermal conductivity of 2.5W/(m·K) has been used in the simulations, instead of the value of 1.2W/(m·K) reported in Table 3. When the property value corresponding to reinforced concrete is considered, the differences with case 2 are larger with respect to the previous case 1.

Table 4. Heat flux transferred by the interior and exterior surfaces and rate of condensed water for the two configurations of thermal bridge.

	Heat Flux (W/m^2)			Rate of Condensed Water $(10^{-5}\ g/s)$
	Interior Blue Surfaces	Interior Green Surfaces	Exterior Pink Surfaces	Whole Structure
Thermal bridge (case 1)	−14.1	−6.3	7.7	3.31
Thermal bridge (case 1.b)	−16.5	−7.4	9.9	4.28
Thermal bridge (case 2)	−14.9	−7.3	8.5	3.71
Percentage difference 1–2	5.7%	16%	10%	12%
Percentage difference 1.b–2	−9.7%	−1.4%	−14%	−13%

Moreover, from the analysis of Figure 7, which shows the temperature contours for the two thermal bridge configurations considered, it is clear that the temperature distribution in the thermal bridge is different in case 1 and case 2. In fact, the inhomogeneities in the structure cause a temperature decrease, particularly evident on the lower floor (blue color in Figure 3), as shown in Figure 7, on the right. Table 4 reports the rate of water condensing in the thermal bridge, which increases by about 12% when inhomogeneities are taken into account.

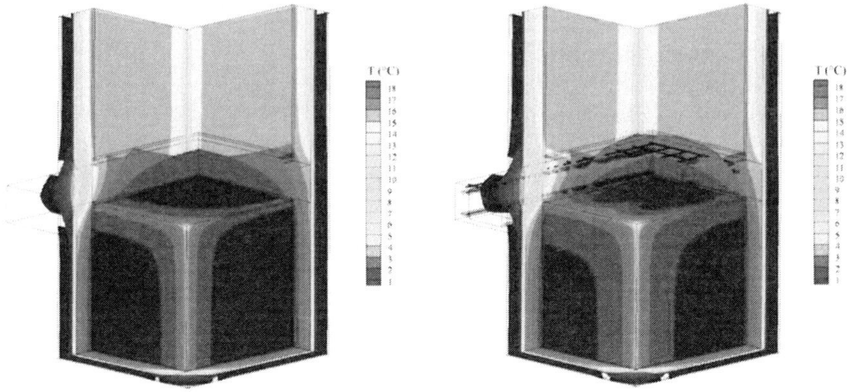

Figure 7. Temperature contours for case 1 (**left**) and case 2 (**right**) under steady state conditions.

6.2. Thermal Bridge under Unsteady Conditions

The two configurations of the thermal bridge, whose 3D profiles are reported in Figure 4 and Figure 5, are also studied under unsteady conditions. In particular, the time-dependent exterior temperature and partial pressure of vapor are considered in the simulations. The vapor partial pressure is calculated on the basis of time-dependent relative humidity, $\phi(\vartheta)$, and time-dependent saturation pressure of water, $psat(T(\vartheta))$, as: $pv(\vartheta)=\phi(\vartheta)psat(T(\vartheta))$. Therefore, the boundary conditions employed for the solution of transient heat transfer are the same of the previous section for the interior and lateral surfaces, while on the exterior surfaces are:

$$-\lambda\frac{\partial T(\vartheta)}{\partial n} = h_{c,e}\left[T_{\infty,e}(\vartheta) - T_{P,e}(\vartheta)\right] \quad \text{with} \quad h_{c,e} = 25\,\mathrm{W}/\left(\mathrm{m}^2\cdot\mathrm{K}\right) \tag{11}$$

The boundary conditions employed for the solution of transient vapor diffusion are represented by Equations (10) for the interior and lateral surfaces, while on the exterior surfaces are:

$$p_{v,e} = p_{v,e}(\vartheta) = \phi_e(\vartheta)\cdot p_{sat}\left(T_{\infty,e}(\vartheta)\right) \tag{12}$$

As initial conditions for both dynamic heat transfer and vapor diffusion, the solution obtained from the steady state simulations has been employed. The values of time-dependent exterior temperature, $T\infty,e(\vartheta)$, and time-dependent relative humidity, $\phi e(\vartheta)$, are calculated for the city of Aosta in Italy on the basis of the technical standard [40], that reports the methodology to be employed for the definition of the reference year to be used to reproduce the climatic conditions. One week hourly simulation has been carried out for the coldest week of the reference year, with temperature values ranging between 9.1 °C and −6.3 °C (Figure 8).

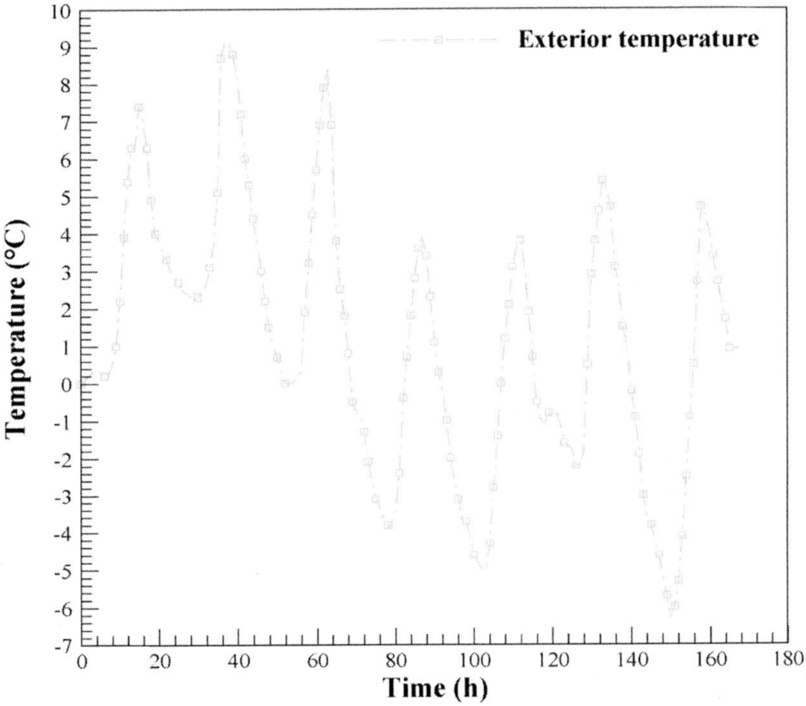

Figure 8. Exterior temperature.

Figure 9 reports the heat flux transferred from the interior surfaces of the two configurations of thermal bridge over the coldest week of the year for the city of Aosta. It is evident that the presence of inhomogeneities causes a different thermal behavior of the structure. In particular, the difference of heat flux between the two cases is almost constant during the week considered in the simulations, since the presence of insulation between the bricks dampens the oscillations of the exterior temperature.

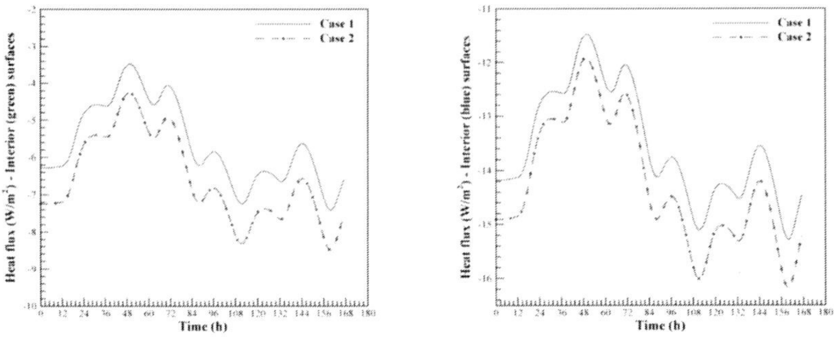

Figure 9. Heat flux transferred from the interior surfaces of the not-heated (**left**) and heated (**right**) environment.

Figure 10 and Figure 11 show the temperature contours obtained for case 1 and case 2, respectively, at four different times (12, 48, 96 and 168 h). From the analysis of these figures, it can be noticed that the temperature distribution in the two configurations of thermal bridge is significantly different at all the time instants presented, due to the inhomogeneities of case 2.

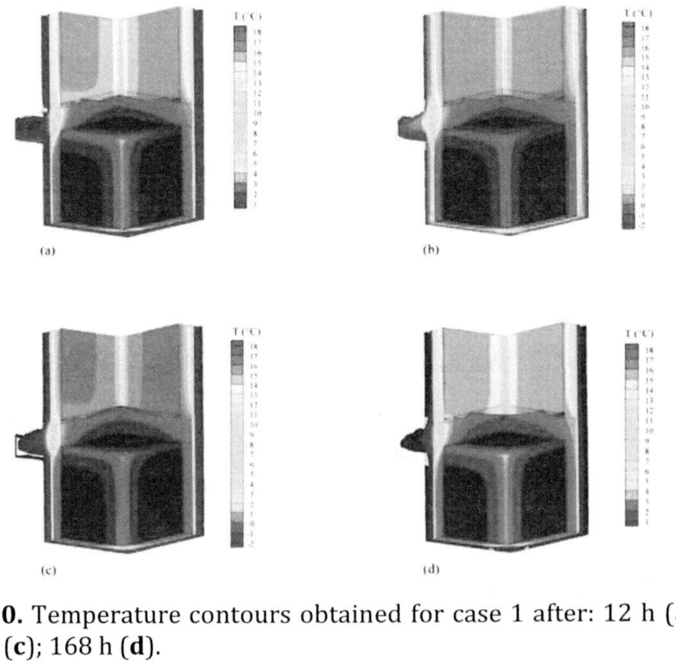

Figure 10. Temperature contours obtained for case 1 after: 12 h (**a**); 48 h (**b**); 96 h (**c**); 168 h (**d**).

Figure 11. Temperature contours obtained for case 2 after: 12 h (**a**); 48 h (**b**); 96 h (**c**); 168 h (**d**).

Figure 12 shows the variation of temperature at point V of Figure 3 for the two cases of thermal bridge considered. This trend confirms the difference in the thermal behavior of the structure, introduced by the metal stud and the brick-concrete slab. Figure 12 also shows the rate of condensed water, $\dot{m}w$, during the considered week. Only positive values of $\dot{m}w$ are reported in Figure 12, meaning that condensation occurs in the structure. From the analysis of the figure, it is clear that the inhomogeneities present in the thermal bridge result in different amounts of condensed water, confirming the differences found in the steady-state simulations.

6.3. Building Envelope under Unsteady Conditions

In order to prove the capability of the present numerical approach to integrate the behavior of 3D thermal bridges in the dynamic simulation of building structures, the geometry shown in Figure 13 has been studied under transient conditions. This geometry represents the room of an apartment located at the top floor of a building, with two exterior walls and the roof exposed to the environment conditions. Two single-glass windows with an aluminum frame have been considered. The steel-concrete pillar studied in the previous section has been employed at the corners of this structure.

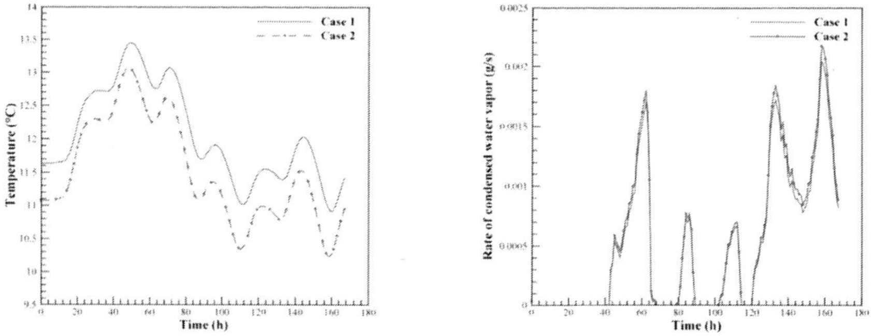

Figure 12. Temperature at point V of Figure 3 (**left**) and rate of condensed water vapor in the structure (**right**) for the two configurations of thermal bridge.

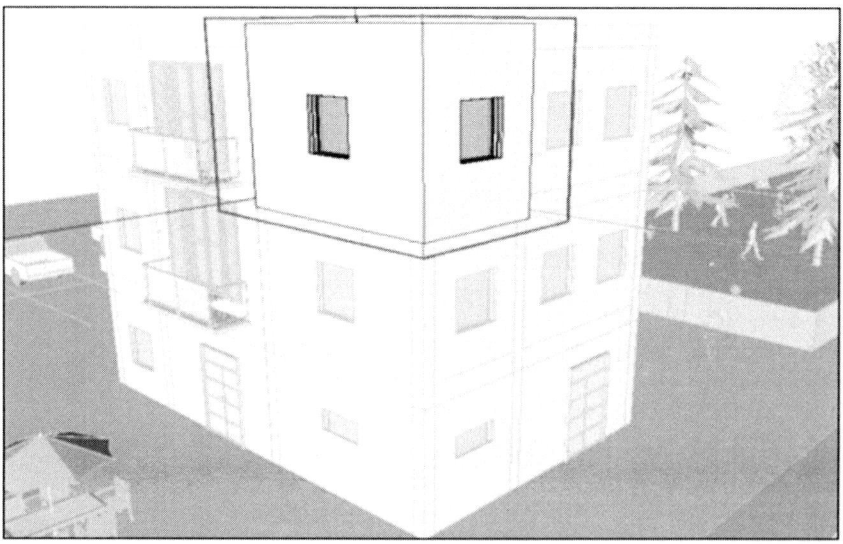

Figure 13. Three-dimensional building envelope.

Figure 14 reports the geometric characteristics of the room envelope taken into account in the simulations. The exterior thermal boundary conditions are represented by Equations (11), including also radiative heat transfer on opaque surfaces, while Equation (12) represents exterior boundary conditions for vapor diffusion. A temperature of 20 °C, with a convective heat transfer coefficient equal to 7.7 W/(m2·K), and a vapor partial pressure of 1168 Pa have been considered as boundary conditions at the interior of the room. Since the whole building has been considered at the same constant interior temperature and relative humidity and, as a consequence, partial pressure of vapor, the vertical wall without window and the pavement have been considered adiabatic and impermeable to mass.

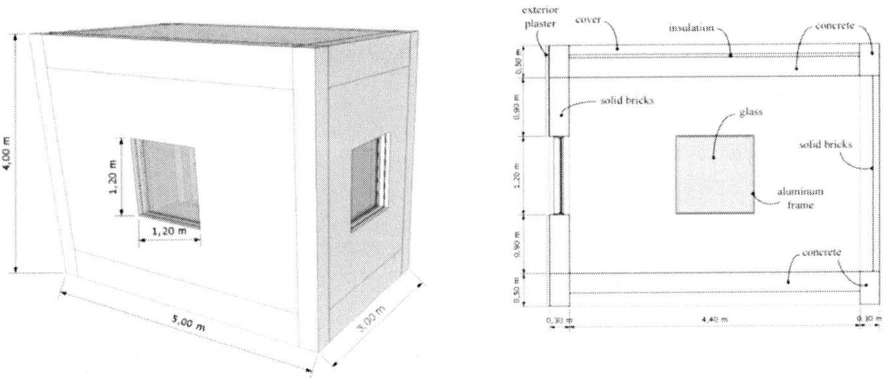

Figure 14. Geometrical characteristics of the room envelope: three-dimensional view (**left**); longitudinal section in correspondence of the window (**right**).

This configuration could be considered representative of the actual behavior of a building envelope. In the authors' opinion, the present 3D microscopic numerical approach should be used in order to evaluate local heat and mass transfer phenomena, while lumped parameter models could be used to reproduce the global behavior of building envelopes. The properties of the solid bricks, insulation material, concrete and plaster are those reported in Table 3. As concerns the other materials used in the simulations carried out in this section, the physical properties, taken from the technical standard [12], are reported in Table 5.

Table 5. Properties of the materials employed in the room envelope.

Material	Thermal Conductivity λ $(W/(m \cdot K))$	Vapor Permeability $\delta \cdot 10^{12}$ $(kg/(m \cdot s \cdot Pa))$	Density	Heat Capacity c $(J/(kg \cdot K))$
Cover	1.5	3.0	2100	1000
Glass	1.0	0.0	2500	750
Aluminum frame	160	0.0	2800	880

Figure 15 shows the temperature contours in the room envelope obtained at four different times (12, 48, 96 and 168 h). From the analysis of this figure, it is evident that the minimum temperatures of the interior side of the envelope are calculated in correspondence of the thermal bridges, and in particular at the corner between the exterior walls.

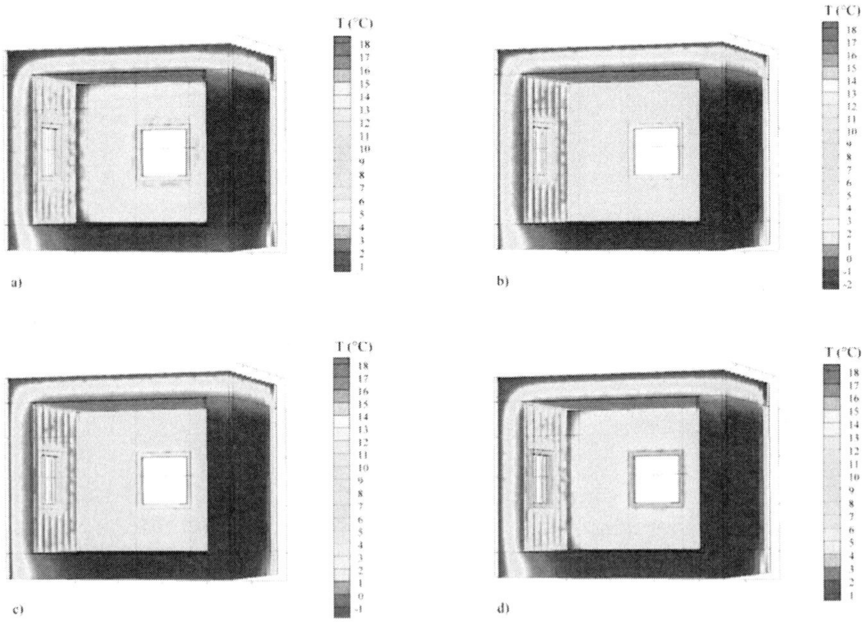

Figure 15. Temperature contours obtained for the room envelope after: 12 h (**a**); 48 h (**b**); 96 h (**c**); 168 h (**d**).

Figure 16 shows the variation of temperature and pressure difference over time at points A and B represented in Figure 17. The results reported in Figure 16 clearly show that the corner between walls, pillar and roof (point

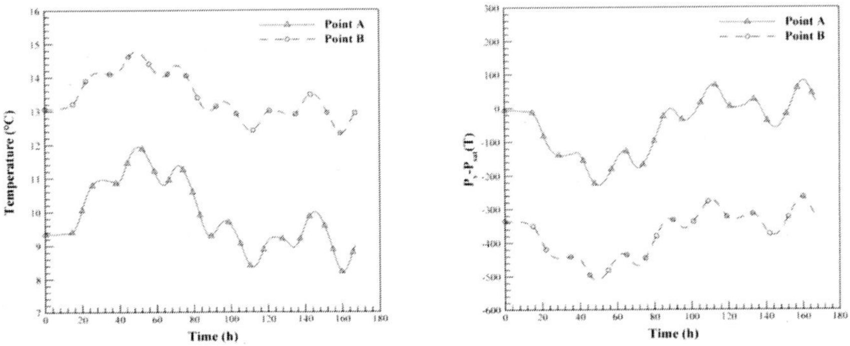

Figure 16. Temperature (**left**) and difference between partial pressure of water vapor and saturation pressure of water (**right**) at points A and B of Figure 17.

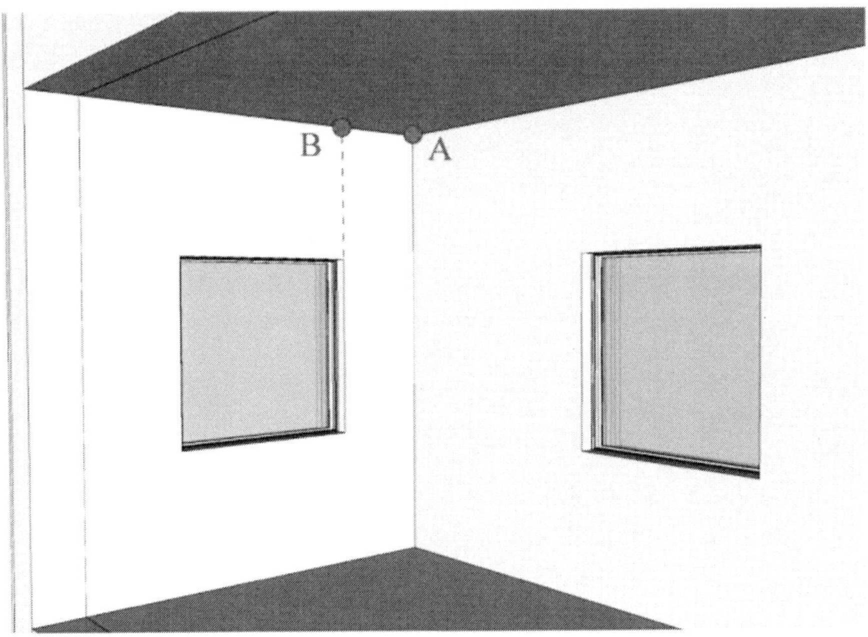

Figure 17. Particular of the room envelope. Point **A** is the corner between walls, pillar and roof, while point **B** is a point at the corner between wall and roof in correspondence of the window's side.

A) is the critical point of the envelope, as expected. In fact, the temperature values at point A are significantly smaller than those of point B during the whole week. As a consequence, the difference between partial pressure of vapor and saturation pressure of water assumes values greater than zero after about 105 h. This means that condensation occurs at the corner (point A), as expected, but not at point B.

7. CONCLUSIONS

In this work, the authors solve steady-state and unsteady conduction heat transfer and vapor diffusion in three-dimensional (3D) inhomogeneous thermal bridges, by using the finite element method. The authors highlight the importance of using a microscopic approach to take into account the inhomogeneities present in structures commonly employed in building envelopes, such as steel-concrete pillars or brick-concrete slabs, in order to estimate heat losses and water condensation in 3D thermal bridges and room envelopes. Both steady-state and unsteady simulations show that the heat flux transferred by 3D thermal bridges and the condensed water can be significantly different if the inhomogeneities in the materials are considered. Steady-state simulations can be useful during a first design stage, while transient ones are required for a proper design or as support to the management and control of heating/cooling systems. The main results obtained in the present paper can be summarized as follows:

The heat flux transferred through the considered 3D thermal bridge assumes a value greater than 16%, when the presence of inhomogeneities in the employed materials is considered by using the present microscopic approach;

The water condensed in the 3D thermal bridge considered is greater than 12%, when the presence of inhomogeneities in the employed materials is considered by using the present microscopic approach;

Critical regions of building envelops, in terms of hygrothermal behavior, can be identified by using the present 3D numerical approach.

The authors think that the use of the present microscopic approach to take into account the inhomogeneities inside construction materials could lead to a reduction of the uncertainties in the estimation of thermal bridges' contribution in building structures. The present microscopic approach allows us to simulate heat and mass transfer in 3D thermal bridges and to highlight critical local phenomena in building envelopes. The present 3D numerical approach can be a useful tool for the energy certification of buildings, especially in their design stage, and could be used together with lumped parameter models that are capable of reproducing the global behavior of entire buildings.

AUTHOR CONTRIBUTIONS

All the authors have contributed to the preparation of this manuscript.

CONFLICTS OF INTEREST

The authors declare no conflict of interest.

REFERENCES

1. Tadeu, A.; Simões, I.; Simões, N.; Prata, J. Simulation of dynamic linear thermal bridges using a boundary element method model in the frequency domain. Energy Build. **2011**, 43, 3685–3695.
2. Energy efficiency: energy Performance of Buildings. Directive 2002/91/EC. 2002.
3. Energy End-Use Efficiency and Energy Services and Repealing Council Directive 93/76/EEC. Directive 2006/32/EC. 2006.
4. Energy Performance of Buildings. Directive 2010/31/EC. 2010.
5. Rendimento Energetico in Edilizia. Directive D.P.R. 02/04/2009 n.59. 2009.
6. Hygrothermal performance of building components and building elements—Internal surface temperature to avoid critical surface humidity and interstitial condensation—Calculation methods. UNI EN ISO 13788:2003. 2003.
7. Theodosiou, T.G.; Papadopoulos, A.M. The impact of thermal bridges on the energy demand of buildings with double brick wall constructions. Energy Build. **2008**, 40, 2083–2089.
8. Thermal bridges in building construction—Linear thermal transmittance—Simplified methods and default values. UNI EN ISO 14683:2001. 2001.
9. Thermal bridges in building construction—Heat flows and surface temperatures—detailed calculation. UNI EN ISO 10211:2008. 2008.
10. Magrini, A.; Magnani, L.; Pernetti, R. The effort to bring existing buildings towards the A class: A discussion on the application of calculation methodologies. Appl. Energy **2012**, 97, 438–450.

11. Materiali da costruzione—Conduttività termica e permeabilità al vapore. UNI 10351:1994. 1994.
12. Building materials and products—Hygrothermal properties—Tabulated design values. UNI EN 12524:2001. 2001.
13. Building materials and products—Hygrothermal properties—Tabulated design values and procedures for determining declared and design thermal values. UNI EN ISO 10456:2008. 2008.
14. Martin, K.; Erkoreka, A.; Flores, I.; Odriozola, M.; Sala, J.M. Problems in the calculation of the thermal bridges in dynamic conditions. Energy Build. **2011**, 43, 529–535.
15. Quinten, J.; Feldheim, V. Dynamic modelling of multidimensional thermal bridges in buildingenvelopes: Review of existing methods, application and new mixed method. Energy Build. **2016**, 110, 284–293.
16. Gao, Y.; Roux, J.J.; Zhao, L.H.; Jiang, Y. Dynamical building simulation: A low order model for thermal bridges losses. Energy Build. **2008**, 40, 2236–2243.
17. Kosny, J.; Christian, J.E. Thermal evaluation of several configurations of insulation and structural materials for some metal stud walls. Energy Build. **1995**, 22, 157–163.
18. Kosny, J.; Kossecka, E. Multi-dimensional heat transfer through complex buildings envelope assemblies in hourly energy simulation programs. Energy Build. **2002**, 34, 445–454.
19. Mao, G.; Johannesson, G. Dynamic calculation of thermal bridges. Energy Build. **1997**, 26, 233–240.
20. Déqué, F.; Ollivier, F.; Roux, J.J. Effect of 2D modelling of thermal bridges on the energy performance of buildings: Numerical application on the Matisse apartment. Energy Build. **2001**, 330, 583–587.
21. He, J.; Hoyano, A.; Asawa, T. A numerical simulation tool for predicting the impact of outdoor thermal environment on building energy performance. Appl. Energy **2009**, 86, 1596–1605.
22. Theodosiou, T.G.; Tsikaloudaki, A.G.; Kontoleon, K.J.; Bikas, D.K. Thermal bridging analysis on cladding systems for building facades. Energy Build. **2015**, 109, 377–384.
23. Nussbaumer, T.; Wakili, K.G.; Tanner, C. Experimental and numerical investigation of the thermal performance of a protected vacuum insulation system applied to a concrete wall. Appl. Energy **2006**, 83, 841–855.
24. Wakili, K.G.; Simmler, H.; Frank, T. Experimental and numerical thermal analysis of a balcony board with integrated glass fibre reinforced polymer GFRP elements. Energy Build. **2007**, 39, 76–81.

25. Zalewski, L.; Lassue, S.; Rousse, D.; Boukhalfa, K. Experimental and numerical characterization of thermal bridges in prefabricated building walls. Energy Convers. Manag. **2010**, 51, 2869–2877.
26. Lorenzati, A.; Fantucci, S.; Capozzoli, A.; Perino, M. Experimental and numerical investigation of thermal bridging effects of jointed Vacuum Insulation Panels. Energy Build. **2016**, 111, 164–175.
27. Mauro, A.; Arpino, F.; Massarotti, N.; Nithiarasu, P. A novel single domain approach for numerical modelling solid oxide fuel cells. Int. J. Numer. Methods Heat Fluid Flow **2010**, 20, 587–612.
28. Mauro, A.; Arpino, F.; Massarotti, N. Three-dimensional simulation of heat and mass transport phenomena in planar SOFCs. Int. J. Hydrogen Energy **2011**, 36, 10288–10301.
29. Arpino, F.; Massarotti, N.; Mauro, A. Efficient three-dimensional FEM based algorithm for the solution of convection in partly porous domains. Int. J. Heat Mass Transf. **2011**, 54, 4495–4506.
30. Carotenuto, A.; Massarotti, N.; Mauro, A. A new methodology for numerical simulation of geothermal down-hole heat exchangers. Appl. Therm. Eng. **2012**, 48, 225–236.
31. Zienkiewicz, O.C.; Taylor, R.L.; Nithiarasu, P. The Finite Element Method for Fluid Dynamics, 6th ed.; Elsevier Butterworth-Heinemann: Oxford, UK, 2005.
32. Lewis, R.W.; Nithiarasu, P.; Seetharamu, K.N. Fundamentals of the Finite Element Method for Heat and Fluid Flow; John Wiley & Sons: Chichester, UK, 2004.
33. Ficker, T. Non-isothermal steady-state diffusion within Glaser's condensation model. Int. J. Heat Mass Transf. **2003**, 46, 5175–5182.
34. Vydra, V. Degradation risk assessment of external envelopes: A practical engineering approach. Build. Environ. **2007**, 42, 344–348.
35. Building components and building elements—Thermal resistance and thermal transmittance—Calculation method. UNI EN ISO 6946:2008. 2008.
36. Regolamento recante norme per la progettazione, l'installazione, l'esercizio e la manutenzione degli impianti termici degli edifici ai fini del contenimento dei consumi di energia. Directive D.P.R. 26/10/1993 n.412. 1993.
37. Energy performance of buildings—Part 1: Evaluation of energy need for space heating and cooling. UNI/TS 11300-1:2008. 2008.
38. Heating systems in buildings—Method for calculation of the design heat load. UNI EN 12831:2006. 2006.

39. Colangelo, F.; Ferone, C.; De Luca, G.; Mauro, A. Experimental and Numerical Analysis of Thermal and Hygrometric Characteristics of Building Structures Employing Recycled Plastic Aggregates and Geopolymer Concrete. Energies **2013**, 6, 6077–6101.
40. Hygrothermal performance of buildings—Calculation and presentation of climatic data Part 4: Hourly data for assessing the annual energy use for heating and cooling. UNI EN ISO 15927-4:2005. 2005.

9

CHAPTER

PERFORMANCE ANALYSIS OF AN EVAPORATOR FOR A DIESEL ENGINE–ORGANIC RANKINE CYCLE (ORC) COMBINED SYSTEM AND INFLUENCE OF PRESSURE DROP ON THE DIESEL ENGINE OPERATING CHARACTERISTICS

Chen Bei [1,2], Hongguang Zhang [1,2,*], Fubin Yang [1,2], Songsong Song [1,2], Enhua Wang [3], Hao Liu [1,2], Ying Chang [1,2], Hongjin Wang [1,2] and Kai Yang [1,2]

[1] *College of Environmental and Energy Engineering, Beijing University of Technology, Pingleyuan No. 100, Beijing 100124, China*

[2] *Collaborative Innovation Center of Electric Vehicles in Beijing, Pingleyuan, No. 100, Beijing 100124, China*

[3] *State Key Laboratory of Automotive Safety and Energy, Tsinghua University, Qinghuayuan, Beijing 100084, China*

ABSTRACT

The main purpose of this research is to analyze the performance of an evaporator for the organic Rankine cycle (ORC) system and discuss the influence of the evaporator on the operating characteristics of diesel engine.

A simulation model of fin-and-tube evaporator of the ORC system is established by using Fluent software. Then, the flow and heat transfer characteristics of the exhaust at the evaporator shell side are obtained, and then the performance of the fin-and-tube evaporator of the ORC system is analyzed based on the field synergy principle. The field synergy angle (β) is the intersection angle between the velocity vector and the temperature gradient. When the absolute values of velocity and temperature gradient are constant and $\beta < 90°$, heat transfer enhancement can be achieved with the decrease of the β. When the absolute values of velocity and temperature gradient are constant and $\beta > 90°$, heat transfer enhancement can be achieved with the increase of the β. Subsequently, the influence of the evaporator of the ORC system on diesel engine performance is studied. A simulation model of the diesel engine is built by using GT–Power software under various operating conditions, and the variation tendency of engine power, torque, and brake specific fuel consumption (BSFC) are obtained. The variation tendency of the power output and BSFC of diesel engine–ORC combined system are obtained when the evaporation pressure ranges from 1.0 MPa to 3.5 MPa. Results show that the field synergy effect for the areas among the tube bundles of the evaporator main body and the field synergy effect for the areas among the fins on the windward side are satisfactory. However, the field synergy effect in the areas among the fins on the leeward side is weak. As a result of the pressure drop caused by the evaporator of the ORC system, the diesel engine power and torque decreases slightly, whereas the BSFC increases slightly with the increase of exhaust back pressure. With the increase of engine speed, power loss, torque loss, and BSFC increment increase gradually, where the most significant change is less than 1%. Compared with the diesel engine itself, the maximum increase of power output of the diesel engine–ORC combined system is 6.5% and the maximum decrease of BSFC is 6.1%.

Keywords: organic Rankine cycle; diesel engine; evaporator; simulation; field synergy; exhaust back pressure

1. INTRODUCTION

Car ownership increases rapidly with economic development. The growing number of cars in use will raise energy consumption and cause serious environmental pollution. Of an engine's fuel combustion energy, only about 30% to 45% is converted into mechanical energy for a diesel engine and about 20% to 30% is converted into mechanical energy for a gasoline engine. About one-third is dissipated in the form of waste heat through the exhaust [1,2]. Therefore, the transformation and utilization of the waste heat of engine exhaust gas is an effective way to improve the efficiency of the engine and reduce fuel consumption and pollutant emissions. Organic Rankine cycle (ORC) is an effective waste heat recovery method that has been widely applied in many domains, such as steel industry [3], cement industry [4], solar-powered plant [5], biomass cogeneration plant [6], internal combustion engine and so on. Uusitalo et al. [7] studied the potential for electricity production from the waste heat of large reciprocating engines with the ORC process technology, which is beneficial for increasing the efficiency of the engine power plant. The results showed that the power output of the selected engine can be increased by 11.4% by utilizing exhaust gas heat and 2.4% by utilizing the charge air heat. Leslie et al. [8] studied a project demonstrating the technical and economic feasibility of recovering thermal energy from a 27 MW gas turbine driving a natural gas pipeline compressor with an organic Rankine cycle system. The results showed that the organic Rankine cycle system produced 5.5 MW of electricity with no additional fuel consumption and near-zero emissions. The organic Rankine cycle improved the overall energy efficiency by 28%.

A key component of the ORC system, the evaporator has an important influence on the thermal power conversion performance of the ORC system and operating performance of the engine. Imran et al. [9] optimized the geometrical parameters of chevron type plate evaporator for a low temperature geothermal ORC system using Non-dominated Sorting Genetic Algorithm-II (NSGA-II). The optimization targets were pressure drop and cost of evaporator. The results showed that the minimum evaporator cost was $1570; the pressure drop was 125 kPa. Zhang et al. [10] investigated the performance of a finned-tube evaporator used to recover diesel engine exhaust waste heat. The heat transfer area of a fin-and-tube evaporator should be chosen according to the typical operating region of the engine. Kundu et al. [11] analyzed the performance and optimum dimensions of flat fins for tube-and-fin heat exchangers through a semi-analytical technique. The optimum dimensions for flat fins of tubes were obtained specifying the ratio of longitudinal to transverse tube pitch. These studies optimize heat exchanger performance through theoretical calculations. However, the flow and heat transfer fields are not discussed.

Several researchers used methods of computational fluid dynamics (CFD) to analyze the flow and heat transfer characteristics of the fluid in the evaporator by establishing a three-dimensional model. Furthermore, guidance for the design of the evaporator was provided. Zhang et al. [12] studied experimentally and numerically the heat transfer characteristics of a helically baffled heat exchanger combined with one three-dimensional finned tube. The results showed that the numerical investigations agreed well with the measurements. Ozden et al. [13] investigated the shell side design of a shell-and-tube heat exchanger by numerically modeling a small heat exchanger. The flow and temperature fields inside the shell were resolved using a commercial CFD package. The effect of the baffle spacing to shell diameter ratio on the heat exchanger performance was investigated. The results showed that CFD simulations could pinpoint the source and the location of the weakness. The use of CFD together with supporting experiments might speed up the shell-and-tube heat exchanger design process. Zhang et al. [14] investigated the performance of several shell-and-tube heat exchangers, one with segmental baffles and four with helical baffles at helix angles of 20°, 30°, 40° and 50°. The results showed that, the heat exchanger with helical baffles of 40° angle showed the best performance among the five heat exchangers that were tested.

Several scholars analyzed the flow and heat transfer characteristics of the tube and fin evaporator through CFD methods. An et al. [15] analyzed the detailed phenomenon in a fin-tube heat exchanger, predicted the heat transfer characteristics of the wave-fin heat exchangers using Fluent, and captured the flow characteristics. Bhuiyan et al. [16] investigated the heat transfer and fluid flow characteristics of a four-row plain fin-and-tube heat exchanger using the CFD simulations. The effects of Reynolds number, fin pitch, and tube pitches on the overall heat transfer and friction factor for plate fin-and-tube heat exchangers were analyzed.

To optimize the structure of the fin and tube evaporator, several scholars analyzed the effect of the geometry of the fin and tube evaporator on heat transfer performance [17,18,19,20,21]. Tang et al. [17] investigated the air-side performance of the fin-and-tube heat exchangers with five types of fin configurations through experimental and numerical methods. The results showed the crimped spiral fin had higher heat transfer and pressure drop than the other four fins. Aytunc et al. [18] investigated the influence of the geometric structure of the fin on heat transfer and pressure drop of heat exchanger based on the numerical simulation method using the CFD software. The results indicated that the distance between the fins had a considerable effect on pressure drop and heat transfer by placing the fin-tube in the downstream region positively. Şahin et al. [19] investigated the heat transfer and pressure drop of seven different fin angles (0°, 5°, 10°, 15°, 20°, 25°, 30°) with plain fin-tube heat exchangers using Fluent. The results showed that, the heat exchanger with the inclined fin angle of 30° exhibited the maximum heat transfer enhancement.

Several scholars studied the performance of the evaporator for an engine exhaust heat recovery system. Hatami et al. [22] separately recycled the exhaust gas energy of gasoline engine and diesel engine by two different evaporators. The evaporators were simulated under three different turbulence models by Fluent software. Then, the variations of the outlet temperatures of the exhaust and the working fluid as functions of the engine torque were obtained. The results showed that the simulation results that selected the SST k-ω and RNG k-ε model were closer to the experimental results. Bai et al. [23] designed six different heat exchangers to recover engine exhaust and compare heat transfer and pressure drop by their CFD models. The results showed that the serial plate structure enhanced heat transfer with seven baffles. The maximum heat was 1737 W, and the maximum pressure drop was 9.7 kPa. However, no study reported flow resistance of the evaporator shell side effect on the engine operation performance.

The evaporator of an ORC system is used for engine exhaust heat recovery, which will cause flow resistance of exhaust gas on the evaporator shell side, thereby resulting in increased exhaust back pressure. As a result, engine performance will be affected.

Presently, quite a few scholars have studied evaporator performance from the point of view of engine waste heat recovery. Furthermore, few scholars study the influence of an evaporator on the engine operating performance comprehensively. Therefore, as a heat exchanging component of the ORC system, the evaporator has key effect on the ORC system performance and the diesel engine performance. The main purpose of this research is to analyze the flow and heat transfer characteristics of the exhaust at the evaporator shell side and the influences of an evaporator used for the ORC system on engine performance under various operating conditions.

First, the evaporator simulation model is established by using Fluent software. The specific boundary conditions are based on the data for the selected engine operating point. The temperature field distribution, the pressure field distribution, and the streamlines of the exhaust gas at the evaporator shell side are obtained through evaporator simulation. The heat transfer performance of the evaporator is analyzed by using the field synergy principle. The test bench of the ORC system for the exhaust heat recovery is built to measure the evaporator performance. The Fluent based simulation results are validated by the experimental data. Furthermore, a simulation model of the diesel engine is built by using GT–Power. The GT–Power based simulation results are validated by the experimental data. The influence of the evaporator for ORC system on engine power, torque, and brake specific fuel consumption (BSFC) is studied under various operating conditions. Finally, the variation tendency of the power output and BSFC of the diesel engine–ORC combined system is obtained.

2. NUMERICAL SIMULATION AND PERFORMANCE ANALYSIS OF EVAPORATOR

2.1. Physical Model of Evaporator

In this research, the flow and heat transfer characteristics of the exhaust at the evaporator shell side are studied by using Fluent software. The half-section physical model of the fin-and-tube evaporator is shown in Figure 1. The evaporator consists of the front end portion, the main body, and the back end portion. The main body is the heat convection place of the exhaust gas and the organic working fluid. The exhaust gas of diesel (hot fluid) is at the evaporator shell side, whereas the organic working fluid (cold fluid) is on the tube side. Both the front and the back end portions are conical pipes, which are used to connect the main body of the evaporator and the exhaust pipes of the diesel engine. The geometric dimensions of the fin-and-tube evaporator are shown in Table 1.

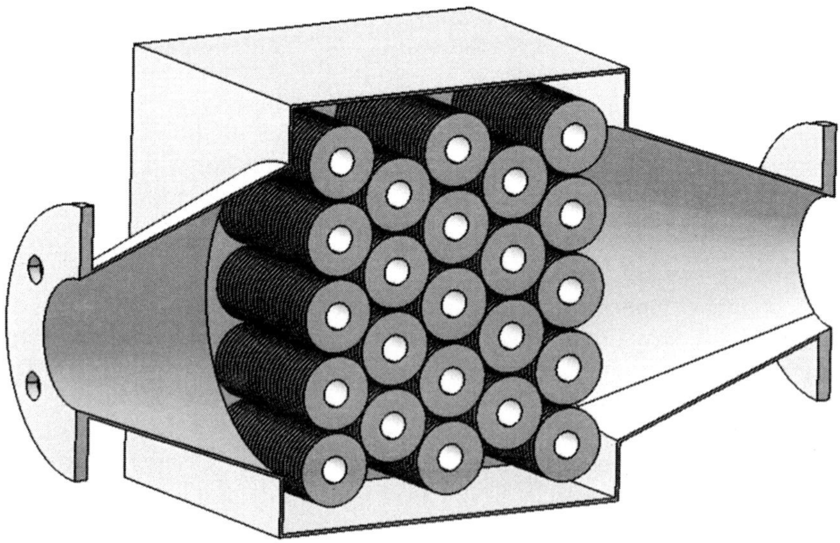

Figure 1. Physical model of the fin-and-tube evaporator.

Table 1. Geometric dimensions of the fin-and-tube evaporator.

Item	Parameter	Unit
Number of tube rows	5	-
Number of tubes in a row	4 or 5	-
Total number of tubes	23	-
Fin thickness	1.5	mm
Fin height	27	mm
Tube outside diameter	27	mm
Tube inner diameter	20	mm
Tube pitch	47 × 56	mm
Tube row alignment	Staggered	-

A quarter of the physical model of the evaporator for the calculations is used to simplify the numerical calculations and ensure the accuracy of the flow and heat transfer calculation. The assumptions are as follows:

(1) The thermal physical properties of the exhaust gas are set as constants based on the average exhaust temperature; (2) The temperatures of fins, round tubes, and shell are set as constants; (3) The flow of exhaust gas at the evaporator shell side is turbulent, and the flow and heat transfer process is steady; (4) The effect of radiative heat transfer, free convection heat transfer, and gravity on the flow of the engine exhaust gas is ignored.

2.2. Boundary Conditions and Governing Equations of the Model

The fin-and-tube evaporator is used to recover the exhaust heat from the diesel engine. The flow and heat transfer characteristics of the engine exhaust at the evaporator shell side at one steady operating point with the engine speed of 2600 $r \cdot min^{-1}$ and the engine torque of 180.7 $N \cdot m$ are studied. The mass flow rate of the intake air and the fuel injection rate are obtained according to the engine test data. The average chemical formula for common fossil diesel fuel can be denoted by $C_{12}H_{23}$ [24]. Based on the atomicity balance principle, the combustion process of diesel fuel with air in the cylinder can be expressed simply as follows:

$$C_{12}H_{23}+17.75\phi_a(O_2+3.773N_2)=12CO_2+11.5H_2O+ 17.75(\phi_a-1)O_2+66.97\phi_a N_2 \quad (1)$$

ϕa (excess air coefficient) is 1.39. Consequently, the mass fractions for components CO_2, H_2O, N_2, and O_2 can be obtained.

Typically, the exhaust gas temperature is high, and the exhaust gas pressure is slightly higher than the atmospheric pressure (at the inlet of the evaporator). Therefore, exhaust gas can be assumed as a mixture of ideal gases. In this study, the thermal physical properties were calculated based on the average temperature of the exhaust gas at the evaporator shell side. The temperature and pressure of engine exhaust gas can be known by experiment. The density and the specific heat at constant pressure, the thermal conductivity, and the dynamic viscosity for CO_2, H_2O, N_2, and O_2 can be computed by using REFPROP software [25].

The average specific heat at constant pressure of the exhaust gas can be calculated as follows:

$$c_{p,m}=\sum_{i=1}^{4} mf_i c_{p,i} \quad (2)$$

The average density of the exhaust gas can be calculated as follows [26]:

$$\rho_m=\frac{\sum_{i=1}^{4} mf_i M_i}{\sum_{i=1}^{4} mf_i M_i / \rho_i} \quad (3)$$

The average dynamic viscosity of the exhaust gas can be calculated as follows [27]:

$$\mu_m=\sum_{i=1}^{4}\frac{vf_i \mu_i}{\sum_{j=1}^{4} vf_j \phi_{ij}} \quad (4)$$

The average thermal conductivity of the exhaust gas can be calculated as follows [27]:

$$\lambda_{\mathrm{m}} = \sum_{i=1}^{4} \frac{\nu f_{\mathrm{i}} \lambda_{\mathrm{i}}}{\sum_{j=1}^{4} \nu f_{\mathrm{j}} \phi_{\mathrm{ij}}} \tag{5}$$

where ϕij can be calculated using the following equation:

$$\phi_{\mathrm{ij}} = \left(\frac{M_{\mathrm{j}}}{M_{\mathrm{i}}} \right)^{1/2} = \phi_{\mathrm{ji}}^{-1} \tag{6}$$

Finally, the thermodynamic properties of the exhaust gas can be obtained, which are listed in Table 2.

Table 2. Thermodynamic properties of the engine exhaust gas.

Item	Parameter	Unit
Exhaust gas density ρ_{m}	0.6553	$\mathrm{kg \cdot m^{-3}}$
Exhaust specific heat at constant pressure $c_{\mathrm{p,m}}$	1096.4	$\mathrm{J \cdot (kg \cdot K)^{-1}}$
Exhaust thermal conductivity λ_{m}	0.04245	$\mathrm{W \cdot (m \cdot K)^{-1}}$
Exhaust dynamic viscosity μ_{m}	2.8177×10^{-5}	$\mathrm{kg \cdot (m \cdot s)^{-1}}$

The boundary of the inlet of the computational domain is set to the "mass flow inlet" (The "mass flow inlet" is a kind of boundary condition which can be used in FLUENT to provide a prescribed mass flow rate or mass flux distribution at the inlet.) in Fluent software. According to the engine test data, the mass flow rate of the intake air is 0.06464 $\mathrm{kg \cdot s^{-1}}$, the fuel injection rate is 0.00324 $\mathrm{kg \cdot s^{-1}}$, and the exhaust temperature at the inlet of the evaporator is 628 K. The mass flow rate of the exhaust is the summation of the intake air rate and the injected fuel rate, which is 0.06788 $\mathrm{kg \cdot s^{-1}}$. The boundary of the outlet of the computational domain is set to the "pressure outlet" (The "pressure outlet" is a kind of boundary condition which can be used in FLUENT to provide a specification of a static pressure at the outlet.) in Fluent software. The exhaust pressure at the outlet of evaporator is set to the ambient pressure (atmospheric pressure). The abovementioned boundary conditions are summarized in Table 3.

Table 3. Boundary conditions of the computational domain.

Item	Parameter	Unit
Mass flow rate of exhaust gas	0.06788	$kg \cdot s^{-1}$
Temperature at evaporator inlet	628	K
Pressure at evaporator outlet	1.01×10^5	Pa

The numerical simulation results based on the standard k-ε, the realizable k-ε, and the RNG k-ε models are compared and discussed. The average exhaust temperature at the evaporator outlet for the standard k-ε model is 488 K, the average exhaust temperature at the evaporator outlet for the RNG k-ε model is 488.4 K, and the average exhaust temperature at the evaporator outlet for the realizable k-ε model is 490.5 K. Among the three models, the average exhaust temperature at the evaporator outlet for the realizable k-ε model is closer to the experimental value. Therefore, the realizable k-ε model is employed during the numerical simulation process. Normally, the standard wall functions in Fluent are used for industrial flows.

2.3. Grid Generation and Numerical Method

The geometry model of the fin-and-tube evaporator is established by using CATIA software. To reduce the workload of calculation, we use a quarter of the evaporator's physical model for calculations based on the principle of symmetry. A 3D grid system is built by using the commercial tool ICEM 14.5. With the use of "Robust (Octree)" mesh method and "tetra/mixed" mesh type, the geometry model is discretized with unstructured elements. The grid of the geometry model for the fin-and-tube evaporator is shown in Figure 2.

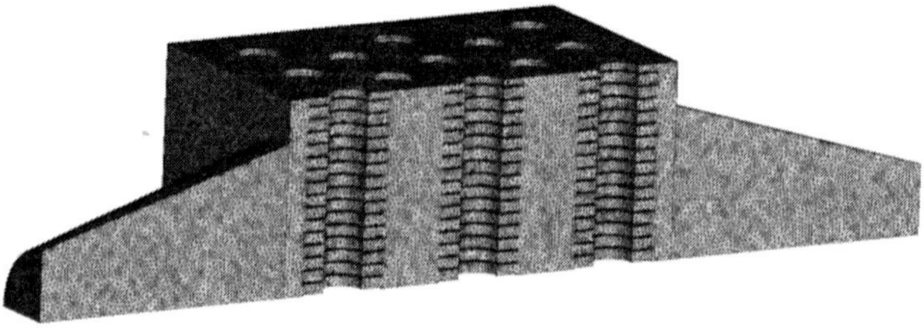

Figure 2. Grid of geometry model for the fin-and-tube evaporator.

To ensure the validity and accuracy of the numerical simulation results, the grid-independence of the numerical solutions should be carefully checked [28]. Three different grid systems are employed. The first grid system has 3,128,845 cells. The second grid system has 4,015,918 cells. Finally, the third grid system has 5,338,761 cells. Through the numerical calculation, the exhaust temperatures at the outlet of the evaporator for the three different grid systems can be obtained. The relative error of the average exhaust temperature at the evaporator outlet between the first grid system and the second grid system is 0.20%. The relative error of the average exhaust temperature at the evaporator outlet between the second grid and the third grid systems is 0.31%. Therefore, the solution precision of the third grid system is satisfactory.

Fluent 14.5 is used for the numerical calculation. A finite volume method is employed to discrete the governing equations. The SIMPLE (semi-implicit method for pressure linked equations) algorithm is adopted during the numerical calculation process. A non-slip boundary condition is assumed for all the solid surfaces of the computational domain. The convergence criteria for the velocity and the momentum equations are the residual errors being less than 10^{-3}, convergence criteria for the energy equation is the residual error being less than 10^{-6}.

2.4. Experimental Validation of Evaporator Numerical Simulation

The diesel engine–ORC combined system mainly includes two parts, diesel engine system and ORC system, in the combined system, the diesel cycle (for the diesel engine) is the topping cycle, and the Rankine cycle (for the ORC system) is the bottoming cycle.

The diesel engine experimental system mainly includes the diesel engine, the measurement and control system, the fuel supply system, the cooling system. The turbocharged diesel engine BJ493ZLQ3 used in this research is a four-cylinder and four-stroke vehicle engine equipped with an electronically controlled high-pressure common-rail fuel injection system.

The ORC experimental system is mainly composed of a pump, evaporator, expander, condenser, reservoir, the measurement and control system for expander, the lubrication system for expander, the cooling system and the data acquisition system for ORC system. Figure 3 shows a schematic diagram of the ORC system used in this study.

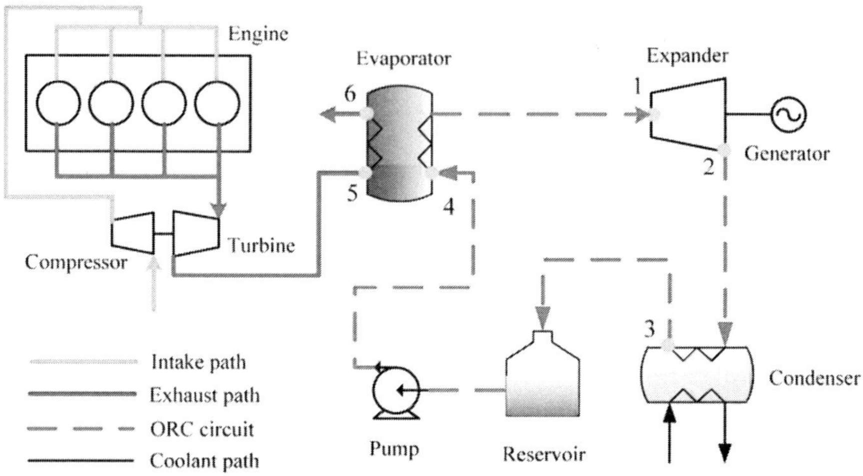

Figure 3. Schematic of the organic Rankine Cycle (ORC) system.

The data acquisition system for ORC system mainly includes the temperature sensors, the pressure sensors, and the flowmeters. The pressure sensors are used for measuring the pressure at the outlet of pump, the outlet of evaporator at the tube side, the outlet of expander, the outlet of condenser. The temperature sensors are used for measuring the temperature at the outlet of pump, the outlet of evaporator at the tube side, the inlet and outlet of the evaporator at the shell side, the outlet of expander, the outlet of condenser. A flowmeter metering the volume flow rate of the working fluid is installed on the pipe between the evaporator and the expander.

During the experiment, the exhaust temperature of the inlet and outlet at the evaporator shell side can be obtained by temperature sensors. Then, we can use the experimental data to validate the numerical simulation results.

When the engine speed is 2600 r·min^{-1} and the engine torque is 180.7 N·m according to the experiment, the measured temperature at the outlet of the evaporator is 491 K. According to the numerical simulation, the average exhaust temperature at the outlet of the evaporator is 490.5 K. The relative error of the exhaust temperature at the evaporator outlet between experimental value and simulation value is 0.1%, which indicates that the numerical simulation correlates well with the actual heat transfer process. During the experiment, several temperature sensors should be installed at the evaporator outlet to obtain the more accurate experimental value, and the average value of the temperature sensors should be taken as the

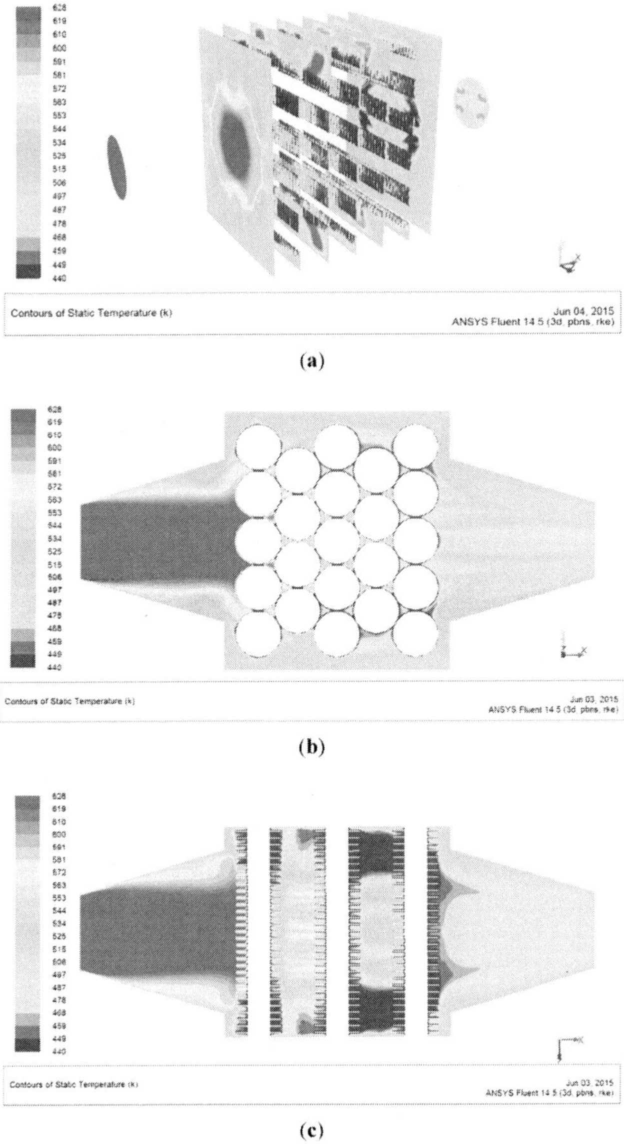

Figure 4. Temperature distribution of exhaust in the evaporator. (**a**) Nine planes in different positions vertical to the x-axis; (**b**) Symmetric plane vertical to the z-axis; (**c**) Symmetric plane vertical to the y-axis.

experimental value. In this research, only one temperature sensor is installed at the evaporator outlet, which can lead to higher relative error of the exhaust temperature at the evaporator outlet between the experimental and simulation values. To validate the numerical results more precisely with the experimental data, more values of the static temperatures of the exhaust should be measured at different positions in the evaporator and later compared with the numerical values.

2.5. Results and Discussion of the Evaporator Numerical Simulation

To reduce the calculation workload, a quarter of the physical model of the evaporator is used for the calculations based on the symmetry principle. Afterwards, the calculation results (for a quarter of the physical model of the evaporator) are symmetrized along the symmetric planes perpendicular to the z-axis and the y-axis. Then, the calculation results for the physical model of the evaporator can be obtained. To analyze the flow and heat transfer characteristics of the engine exhaust at the evaporator shell side, nine planes in different positions vertical to the x-axis are selected, which are shown in Figure 4a. The nine planes include the plane of the inlet of the entire flow domain, the plane of the outlet of the entire flow domain, the interface between the front end portion and the main body, the interface between the back end portion and the main body, and the planes through the centerlines of the tube bundles in each row along the direction of the exhaust flow. In addition, the symmetric plane vertical to the z-axis and symmetric plane vertical to the y-axis are selected, which are shown in Figure 4b,c.

The temperature distribution of the exhaust at the evaporator shell side is shown in Figure 4. The exhaust temperature in the front end portion approximately equals the inlet exhaust temperature of 628 K. In addition, the flow areas of the high temperature exhaust gas are nearly round, and expand gradually along the flow direction. The main reason is that the shape of the front end portion that connects with the main body is round, and the exhaust gas temperatures decrease gradually along the flow direction. The high-temperature exhaust gas in the flow space between the first and second rows diffuses quickly.

The pressure distribution of the exhaust at the evaporator shell side is shown in Figure 5. As shown in Figure 5, the maximum pressure exists in the central areas near the tube bundles of the first row towards the front end portion. The high temperature exhaust in the conical divergent pipe is blocked by the tube bundles of the first row; the exhaust mass flow rate decreases rapidly, then the exhaust pressure increases rapidly. The maximum pressure is about 169 Pa (relative pressure) and is significantly lower than the maximum limit of 30 kPa. The pressure drop of the exhaust gas between the inlet and outlet of the evaporator is about 78 Pa, which is the exhaust back pressure increment at this operating point of the diesel engine.

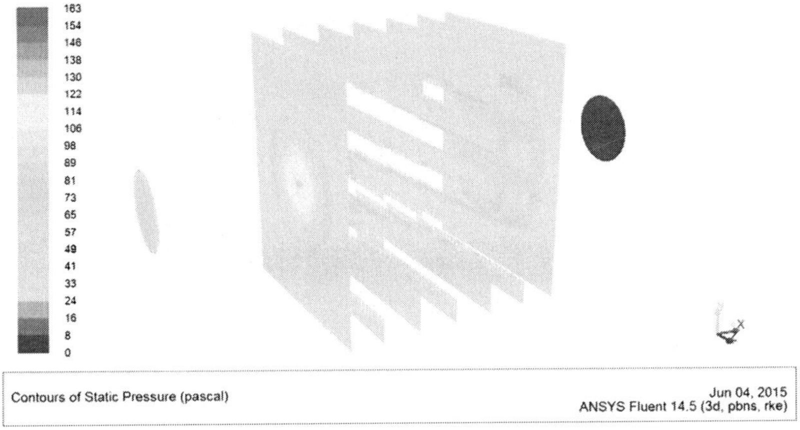

(a)

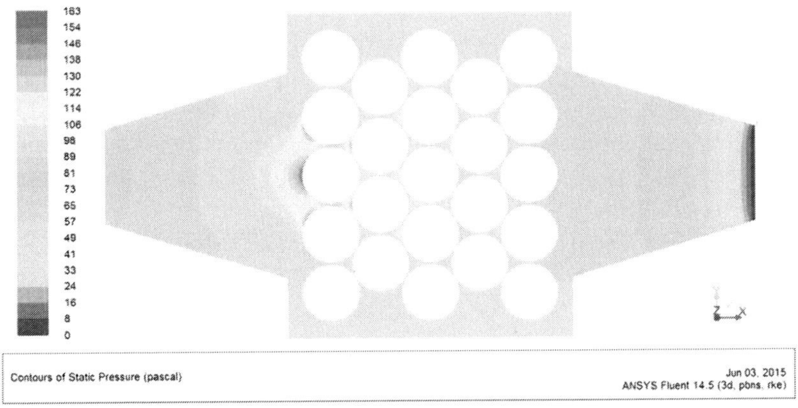

(b)

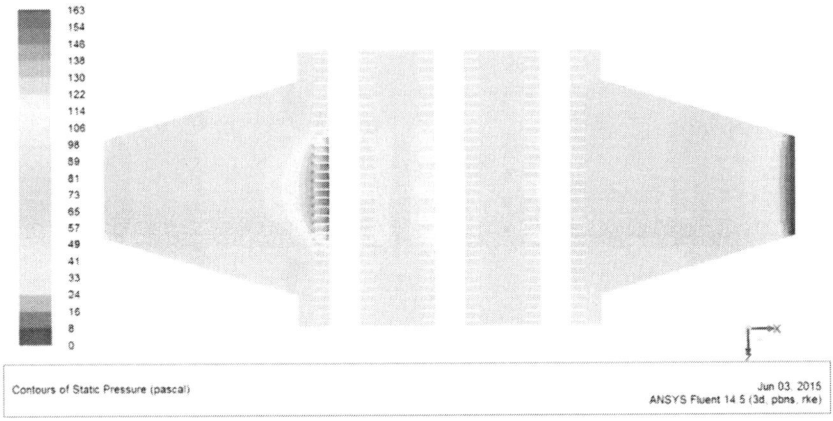

(c)

Figure 5. Pressure distribution of exhaust in the evaporator. (**a**) Nine planes in different positions vertical to the x-axis; (**b**) Symmetric plane vertical to the z-axis; (**c**) Symmetric plane vertical to the y-axis.

The streamlines of the exhaust gas in the evaporator are shown in Figure 6. Figure 6a is the view in the negative direction of the z-axis. Figure 6b is the view in the negative direction of the y-axis. In the central space of the evaporator, the exhaust flow is well distributed, which means the convective heat transfer is fully developed. In the four corners far from the center of the evaporator and near the front end portion, almost no exhaust flow exists, which means that the convective heat transfer is very weak. The exhaust flow that enters the main body is restricted in the domains close to the central space of the evaporator because the shape of the front end portion that connects with the main body is round. Therefore, to ensure the exhaust flow is evenly distributed in the spaces between the neighboring fin layers, a reasonable design of the shapes that connect the main body with the front end and back end portions is necessary.

To improve the heat transfer performance of the fin-and-tube evaporator from the viewpoint of the field synergy principle, the field synergy angle between the velocity and the temperature gradient in the flow field is discussed. The field synergy principle is different from the traditional heat transfer enhancement theory [29,30,31]. According to the field synergy principle, the convective heat transfer performance not only depends on the velocity of the fluid, the thermodynamic properties of the fluid, and the temperature difference between the fluid and solid wall, but also on the situation of synergy between the velocity field and the temperature gradient field.

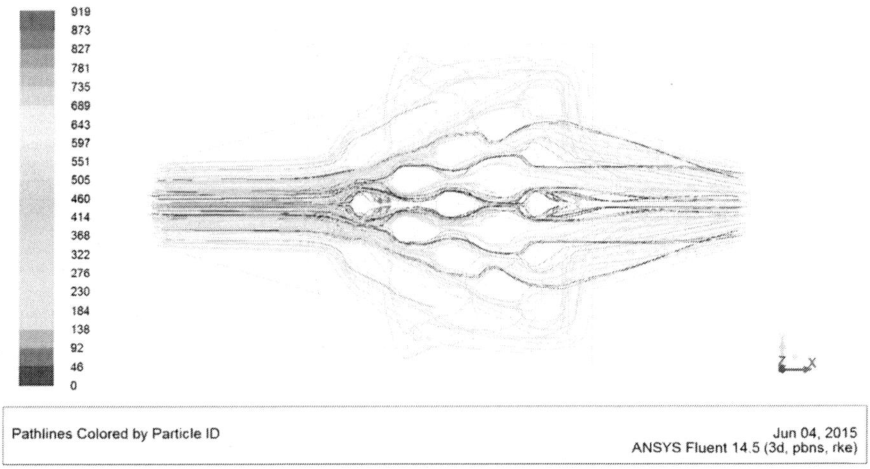

(a)

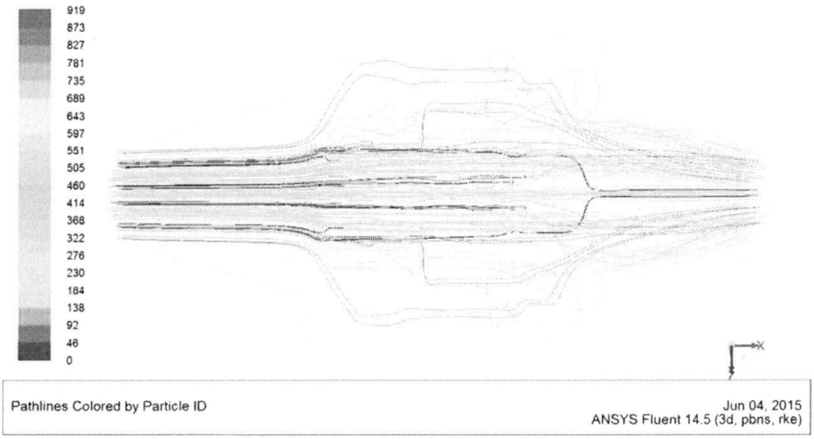

(b)

Figure 6. Streamlines of the exhaust gas in the evaporator. (**a**) The view in the negative direction of the z-axis; (**b**) The view in the negative direction of the y-axis.

For a steady, 2D boundary layer flow over a cold flat plate at zero incident angle, as shown in Figure 7a. The energy equation is:

$$\rho c_{\mathrm{p}}\left(u\frac{\partial T}{\partial x}+v\frac{\partial T}{\partial y}\right)=\frac{\partial}{\partial y}\left(\lambda\frac{\partial T}{\partial y}\right) \tag{7}$$

The energy equation for conduction with the heat source between two parallel plates at constant but different temperatures as shown in Figure 7b is:

$$-\dot{q}=\frac{\partial}{\partial y}\left(\lambda\frac{\partial T}{\partial y}\right) \tag{8}$$

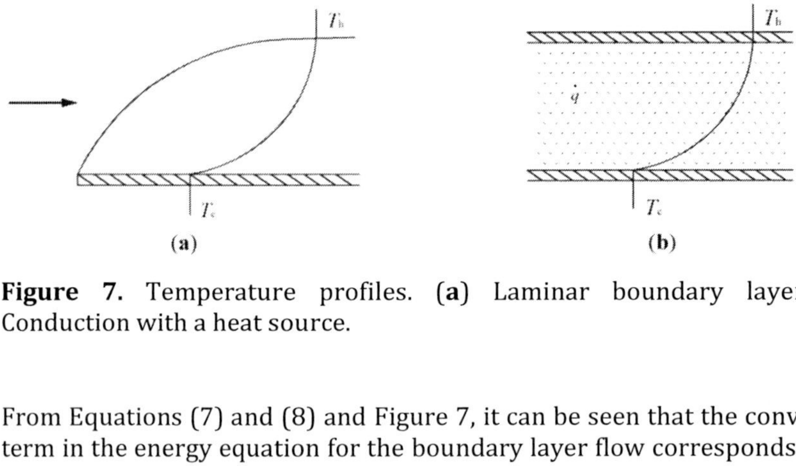

Figure 7. Temperature profiles. (**a**) Laminar boundary layer; (**b**) Conduction with a heat source.

From Equations (7) and (8) and Figure 7, it can be seen that the convection term in the energy equation for the boundary layer flow corresponds to the heat source term in the conduction equation. The presence of heat sources leads to increased heat flow at the boundary for both the conduction and convection problems. The integral of Equation (7) over the thickness of the thermal boundary layer is:

$$\int_0^{\delta_t,R}\rho c_{\mathrm{p}}\left(u\frac{\partial T}{\partial x}+v\frac{\partial T}{\partial y}\right)\mathrm{d}y=-\lambda\left.\frac{\partial T}{\partial y}\right|_{w} \tag{9}$$

From Equation (9), we can see that the wall heat flow is equal to the overall strength of the heat sources inside the thermal boundary layer. This means that the convection heat transfer can be enhanced by increasing the quantity of the integral of the convection terms over the thermal boundary layer. Equation (9) can be rewritten with the convection term in vector form as:

$$\int_0^{\delta_t, R} \rho c_p (U \cdot \nabla T) \mathrm{d}y = -\lambda \frac{\partial T}{\partial y}\bigg|_w \quad (10)$$

According to Equation (10), for a certain flow rate and temperature difference between the wall and the incoming flow, the wall heat flow increases with the decrease of the included intersection angle between the velocity and temperature gradient.

With the following dimensionless variables for the boundary layer flow:

$$\overline{U} = \frac{U}{U_\infty}, \quad \nabla\overline{T} = \frac{\nabla T}{(T_\infty - T_w)/\delta_t}, \quad \overline{y} = \frac{y}{\delta_t}, \quad T_\infty > T_w \quad (11)$$

Equation (10) can be written in the dimensionless form:

$$Nu_x = Re_x Pr \int_0^1 (\overline{U} \cdot \nabla\overline{T}) \mathrm{d}\overline{y} \quad (12)$$

From Equation (12), it can be seen that there are two ways to enhance heat transfer: (a) increasing Reynolds or/and Prandtl number; which is well known in the literature; (b) increasing the value of the dimensionless integration.

The vector dot product, $\overline{U} \cdot \nabla\overline{T}$ in the dimensionless integration in Equation (12) can be expressed as:

$$\overline{U} \cdot \nabla\overline{T} = \left|\overline{U}\right| \cdot \left|\nabla\overline{T}\right| \cos\beta \quad (13)$$

β is the field synergy angle. The field synergy angle is the intersection angle between the velocity vector and the temperature gradient.

Deduced from Equation (12) is:

$$St = \frac{Nu}{Re_x Pr} = I \quad (14)$$

$$I = \int_0^1 (\overline{U} \cdot \nabla\overline{T}) d\overline{y} \quad (15)$$

$$I = f(Re_x, Pr) \tag{16}$$

According to the abovementioned equations, we can see that convective heat transfer performance is significantly relative to the situation of synergy between the velocity field and the temperature gradient field. From the point of view of scalar, the synergy between the velocity field and the temperature gradient field is the synergy among the absolute value of velocity, the absolute value of temperature gradient, and the cosine of the intersection angle between the velocity vector and the temperature gradient (β, field synergy angle). When the absolute values of velocity and temperature gradient are constant and $\beta < 90°$, I increases with the decrease of the β. Thus, Nu increases and heat transfer enhancement can be achieved. When the absolute values of velocity and temperature gradient are constant and $\beta > 90°$, I increases with the increase of the β. Thus, Nu increases and heat transfer enhancement can be achieved. Thus, coordinating the velocity field and the temperature gradient field is a novel way for heat transfer enhancement.

Previously, we studied the synergy situation of a fin-and-tube evaporator [32]. However, in that study, although the velocity vectors and the temperature gradient vectors were obtained, detailed synergy angles were not determined. In this study, a program is developed in CFD-post software to calculate the synergy angles at various positions of the evaporator shell side. The calculation results are shown in Figure 8. The synergy angle distribution for nine planes in different positions vertical to the x-axis is shown in Figure 8a. The synergy angle distribution for the symmetric plane vertical to the z-axis and y-axis is shown in Figure 8b and Figure 8c, respectively. The blue and red areas in Figure 8 represent the synergy angles between the velocity and temperature gradient are 0° and 180°, respectively. Correspondingly, the effect of the field synergy between the velocity and the temperature gradient is satisfactory, then heat transfer enhancement can be achieved. The green area in Figure 8 represents the synergy angle between the velocity and temperature gradient is 90°, where the field synergy between the velocity and temperature gradient is weak.

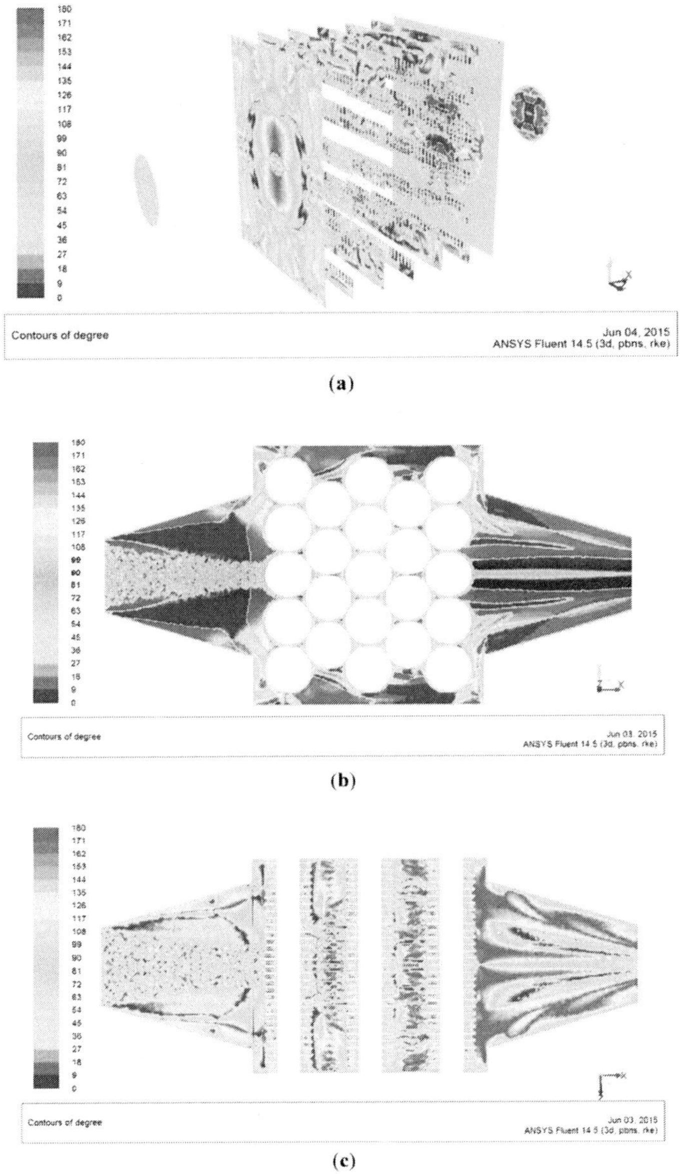

Figure 8. Synergy angle distribution of exhaust in the evaporator. (**a**) Nine planes in different positions vertical to the x-axis; (**b**) Symmetric plane vertical to the z-axis; (**c**) Symmetric plane vertical to the y-axis.

As we can see from Figure 8, the synergy angle between the velocity and temperature gradient are close to the blue or red color (0° or 180°) in the areas among the tube bundles of the evaporator main body. Thus, the field synergy effect for these areas is satisfactory. The synergy angle between the velocity and the temperature gradient is close to the red (180°) in the areas among the fins on the windward side. Thus, the field synergy effect for these areas is satisfactory. However, the synergy angle is close to the green (90°) in the areas among the fins on the leeward side. Thus, the field synergy effect for these areas is weak. The temperature gradients in the areas close to the fins are nearly vertical to the surface of the fins, which is also vertical to the flow direction of the exhaust, because of the relatively large external diameters of the fins and the heat conduction effects. To further improve heat transfer performance, the layout of the fins can be modified, and the shapes and angles of the fins can be adjusted to make the synergy angle between the velocity and the temperature gradient close to 0° or 180°.

3. NUMERICAL SIMULATION FOR DIESEL ENGINE

3.1. Simulation Model of Diesel Engine

In this research, the turbocharged diesel engine BJ493ZLQ3 is a four-cylinder and four-stroke vehicle engine equipped with an electronically controlled high-pressure common-rail fuel injection system. The simulation model for BJ493ZLQ3 diesel engine is built by using GT–Power software, which is shown in Figure 9. The simulation model mainly includes an air intake system, four fuel injectors, four cylinders, a crankcase, a turbocharger, and an exhaust system. The main technical parameters of the BJ493ZLQ3 diesel engine are listed in Table 4.

3.2. Validation of Diesel Engine Simulation Model

In order to ensure that the simulation model can accurately describe the actual operation of the diesel engine and effectively predict the performance of the diesel engine, the simulation results for various operating points are verified by using the test data of BJ493ZLQ3 diesel engine.

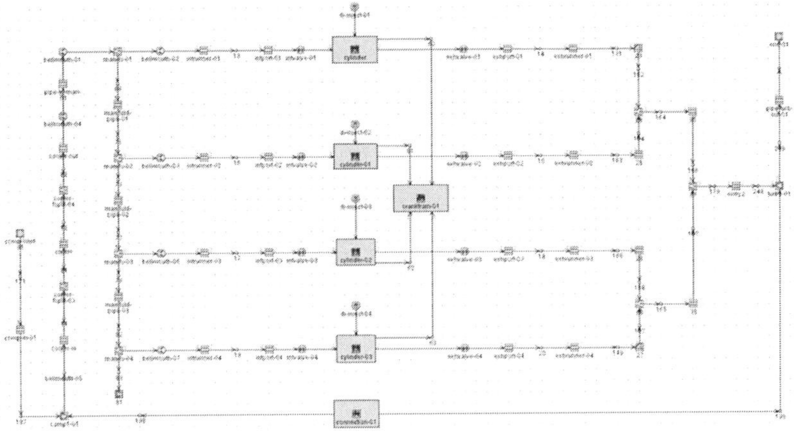

Figure 9. Simulation model of BJ493ZLQ3 diesel engine.

Table 4. Technical parameters of the BJ493ZLQ3 diesel engine.

Item	Parameter	Unit
Type	turbocharged common-rail diesel engine	-
Cylinder diameter	93	mm
Stroke	102	mm
Compression ratio	17.4	-
Displacement	2.771	L
Rated power	85	kW
Rated speed	3600	$r\cdot min^{-1}$
Maximum torque	280	N·m
Speed at max. torque	2300	$r\cdot min^{-1}$

The simulation and experiment results of engine power at full load are shown in Figure 10. Simulation results have the same variation trend with experiment results. When the diesel engine speed is 3400 $r\cdot min^{-1}$, the relative error between the simulation and experiment values of the engine power reaches its maximum of approximately 4.55%.

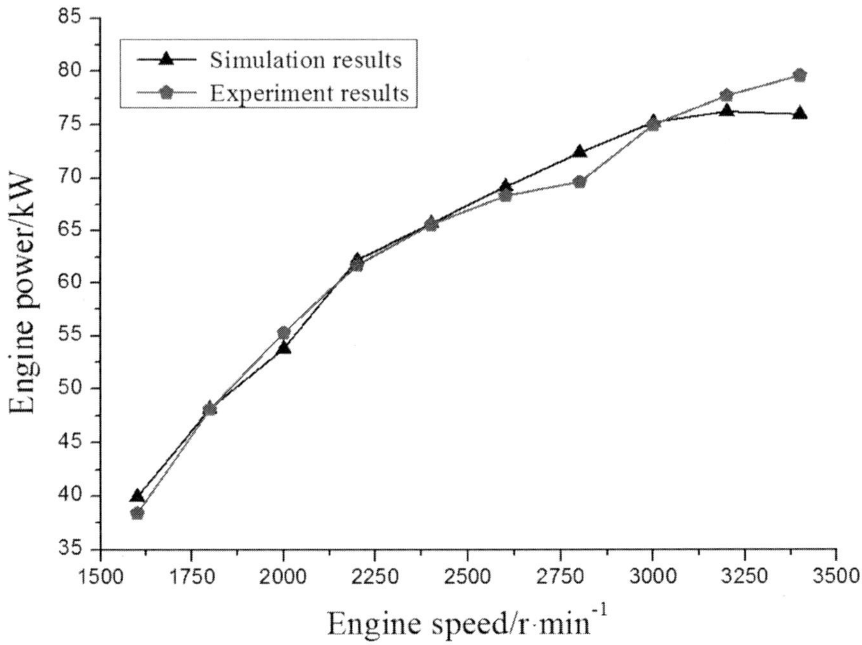

Figure 10. Simulation and experiment results of engine power at full load.

The simulation and experiment results of the engine torque at full load are shown in Figure 11. Simulation results have the same variation trend with experiment results. When the diesel engine speed is 3400 r·min^{-1}, the relative error between the simulation and experiment values of the engine torque reaches its maximum of approximately 4.65%.

The simulation and experiment results of the BSFC of BJ493ZLQ3 diesel engine at full load are shown in Figure 12. Simulation results have the same variation trend with experiment results. When the diesel engine speed is 3400 r·min^{-1}, the relative error between the simulation and experiment values of the BSFC reaches its maximum of approximately 4.97%.

According to the abovementioned analysis, the maximum relative error between the simulation and the experiment values is less than 5%. Therefore, it can be concluded that the engine simulation model can be employed with high precision.

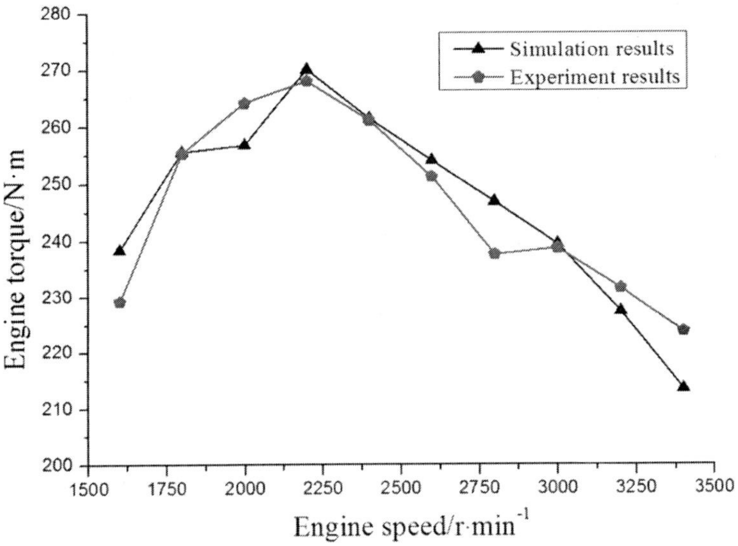

Figure 11. Simulation and experiment results of engine torque at full load.

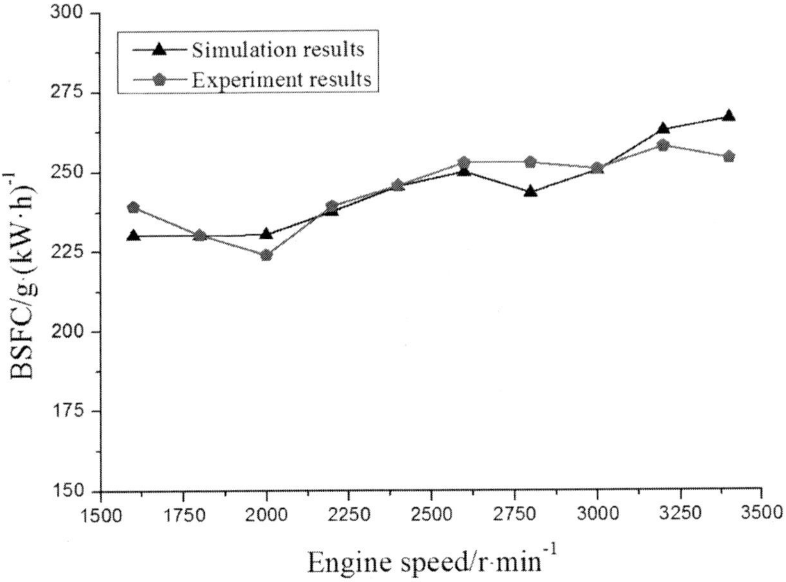

Figure 12. Simulation and experiment results of brake specific fuel consumption (BSFC) at full load.

4. ANALYSIS OF THE OPERATING PERFORMANCE FOR DIESEL ENGINE–ORC COMBINED SYSTEM

4.1. Influence of Pressure Drop of Evaporator on the Engine Operating Characteristics

In this study, the influence of the pressure drop of the evaporator of the ORC system on engine operating characteristics is analyzed for 14 different operating points at full load (the engine speed is selected as 1000, 1200, 1400, 1600, 1800, 2000, 2200, 2400, 2600, 2800, 3000, 3200, 3400, and 3600 r·min^{-1}, respectively). According to the pressure drop between the inlet and the outlet of the evaporator based on numerical simulation results of the evaporator, the exhaust back pressure increment of the diesel engine is obtained. When the diesel engine is combined with the ORC system, the diesel engine–ORC combined system is established, which is beneficial in improving fuel economy and power performance. Considering the diesel engine–ORC combined system, the variation of the engine power, engine torque, and BSFC is discussed under various operating points by the simulation based on GT–Power software.

The exhaust back pressure increment of the diesel engine at full load with consideration of the diesel engine–ORC combined system is shown in Figure 13. With the increase of engine speed, the exhaust back pressure increment increases gradually because the intake air mass flow rate of the diesel engine gradually increases with the increase of the engine speed, which results in the increase of the engine exhaust mass flow rate. The pressure drop of the engine exhaust increases at the evaporator shell side.

With consideration of the diesel engine–ORC combined system, the power loss, the torque loss, and the BSFC increment of the diesel engine are discussed under various operating points and shown in Figure 14, Figure 15 and Figure 16. Both the engine power and torque decrease, with a maximum decrease of less than 1%. The BSFC increases with a maximum increase of less than 1% because the flow resistance of the engine exhaust on the evaporator shell side exists, thereby increasing the exhaust back pressure of the diesel engine.

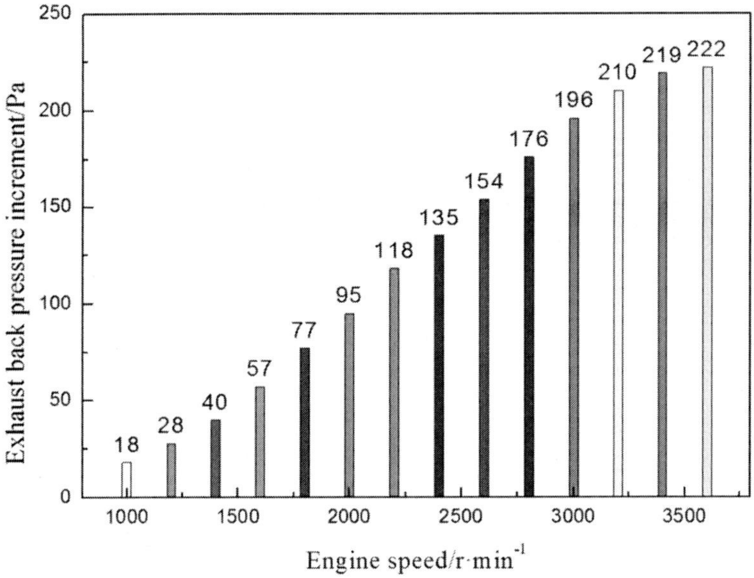

Figure 13. Exhaust back pressure increment of diesel engine at full load.

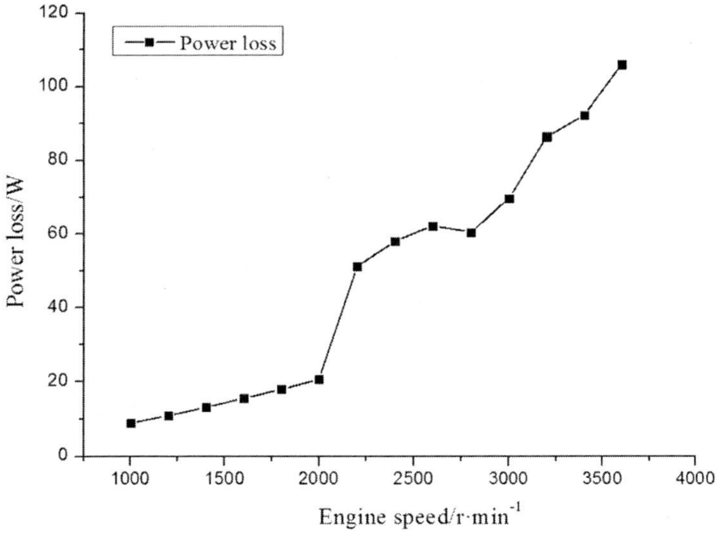

Figure 14. Power loss of diesel engine at full load.

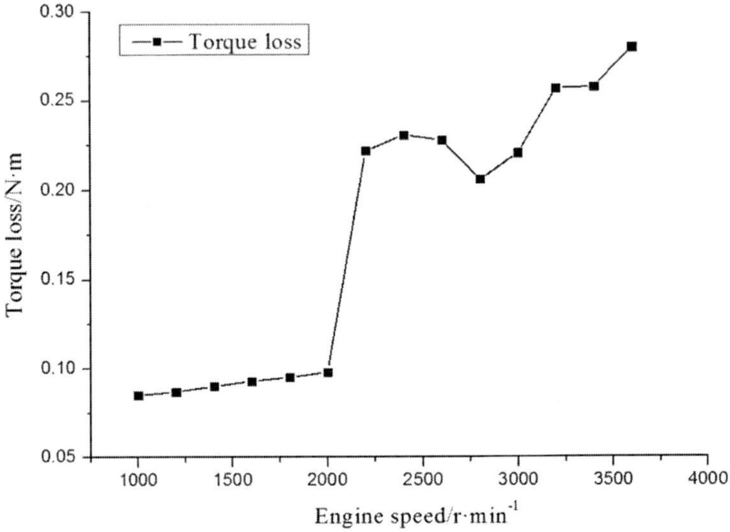

Figure 15. Torque loss of diesel engine at full load.

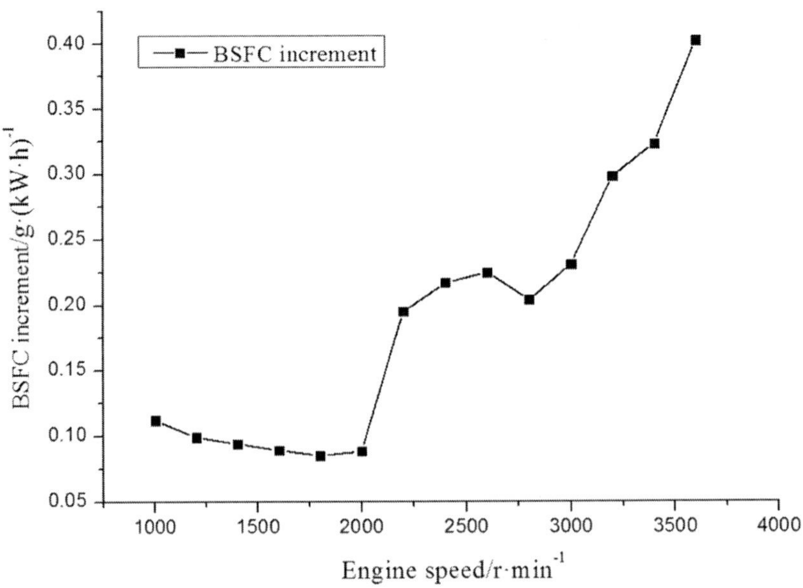

Figure 16. BSFC increment of diesel engine at full load.

4.2. Thermodynamic Model of the ORC System

The T-s diagram of the ORC system is shown in Figure 17. The ORC system comprises a pump, an evaporator, an expander, a condenser, and a reservoir. Process 5–6 is the heat rejection process of the diesel engine exhaust at the evaporator shell side. Process 1–2 is the actual expansion process, whereas Process 1–2s is the isentropic expansion process in which the working fluid soon turns into low-temperature and low-pressure gas. At the same time, the expander produces useful work used to generate electricity. Process 2–3 is the constant pressure heat rejection process, in which the working fluid turns into liquid in the condenser and flows into the reservoir. Process 3–4 is the actual compression process, whereas Process 3–4s is the isentropic compression process in which the working fluid is pressurized into high-pressure liquid state by using a pump. Process 4–1 is the constant pressure heat addition process in which the working fluid absorbs the heat from the engine exhaust in the evaporator and soon turns into high-temperature and high-pressure gas.

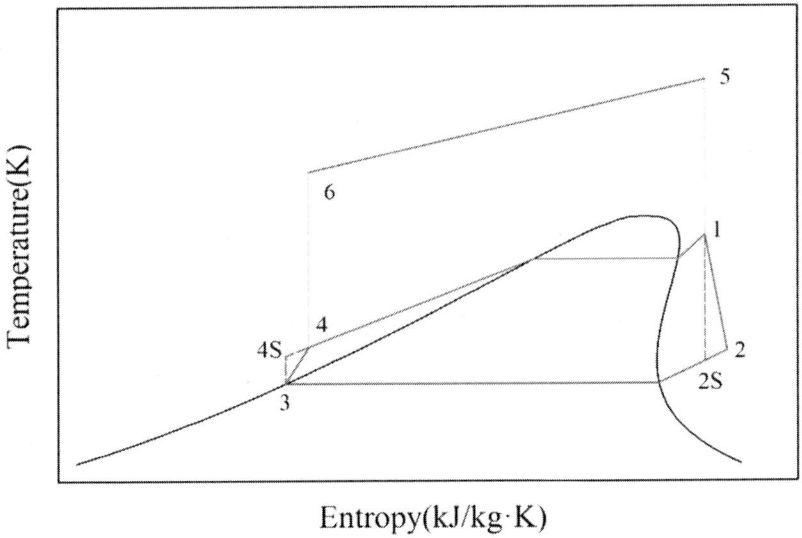

Figure 17. T-s diagram of the ORC system.

The power output of the expander Ws can be calculated as follows:

$$\dot{W}_s = \dot{m}\left(h_1 - h_2\right)\eta_s \tag{17}$$

The heat transfer rate of the condenser $Q^{\cdot}c$ can be calculated as follows:

$$\dot{Q}_c = \dot{m}\left(h_2 - h_3\right) \tag{18}$$

The power consumed by pump $W^{\cdot}p$ can be calculated as follows:

$$\dot{W}_p = \dot{m}\left(h_4 - h_3\right) = \dot{m}\left(h_{4s} - h_3\right) / \eta_p \tag{19}$$

The heat transfer rate of the evaporator $Q^{\cdot}e$ can be calculated as follows:

$$\dot{Q}_e = \dot{m}\left(h_1 - h_4\right) \tag{20}$$

The net power output of ORC system $W^{\cdot}n$ can be calculated as follows:

$$\dot{W}_n = \dot{W}_s - \dot{W}_p \tag{21}$$

In this study, the following assumptions are made for the ORC system:

1) The pressure drop and heat loss in each component as well as pipelines are neglected.
2) The evaporation pressure ranges from 1.0 MPa to 3.5 MPa.
3) The ambient temperature is set to 293 K.
4) The expansion ratio of the expander is limited to 8.
5) The working fluid releases heat in the condenser and then turns into saturated liquid.
6) The isentropic efficiencies of the expander and the pump are both set to 0.8.

To evaluate the performance of the diesel engine–ORC combined system, the power increase rate of the diesel engine–ORC combined system can be expressed as

$$\eta_w = \frac{\dot{W}_n + \dot{W}_b - \dot{W}_i}{\dot{W}_i} \times 100\% \tag{22}$$

The BSFC of the diesel engine–ORC combined system can be expressed as

$$bsfc_{\text{com}} = \frac{\dot{m}_{\text{fuel}}}{\dot{W}_{\text{n}} + \dot{W}_{\text{b}}} \times 3.6 \times 10^{6} \tag{23}$$

The BSFC decrease rate of the diesel engine–ORC combined system can be expressed as

$$\eta_{\text{bsfc}} = \frac{bsfc - bsfc_{\text{com}}}{bsfc} \times 100\% \tag{24}$$

4.3. Operating Performance of Diesel Engine–ORC Combined System

In this research, R245fa is selected as the working fluid. Figure 18 shows the variation of the power increase rate of the diesel engine–ORC combined when the evaporation pressure ranges from 1.0 MPa to 3.5 MPa for the 14 operating points at full load of the diesel engine. The variation of the BSFC decrease rate of the diesel engine–ORC combined system is shown in Figure 19.

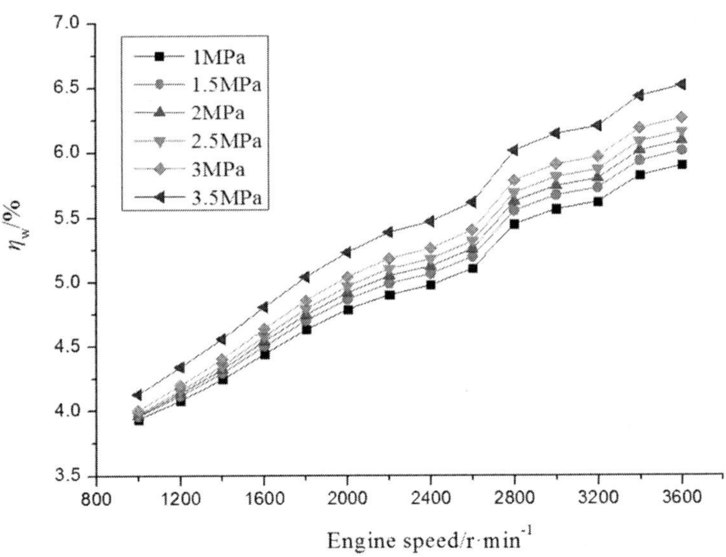

Figure 18. Variation of ηw with engine speed and evaporation pressure.

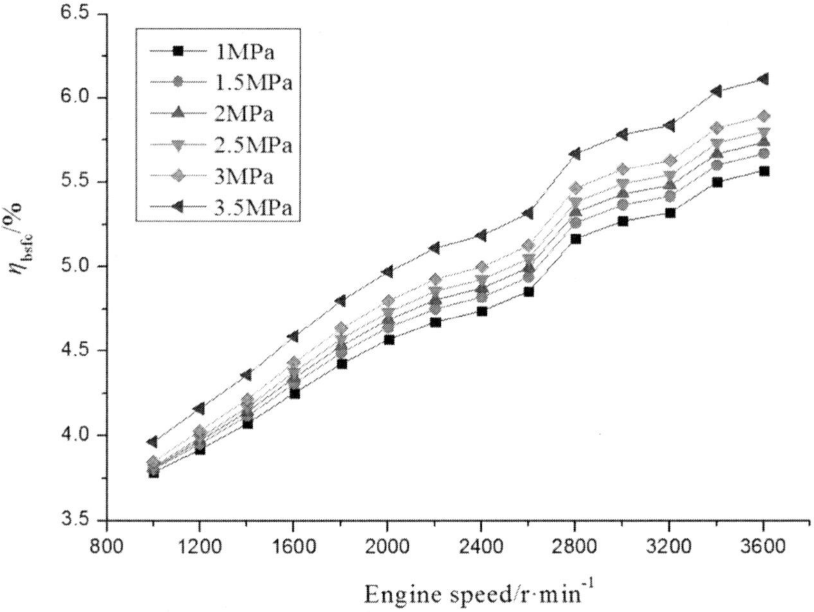

Figure 19. Variation of ηbsfc with engine speed and evaporation pressure.

As shown in Figure 18, when the engine speed is kept constant, the power increase rate of the diesel engine–ORC combined system gradually increases with the increase of evaporation pressure. The main reason is that, on the one hand, the net power output of the ORC system increases with the increase of evaporation pressure. On the other hand, the power loss of diesel engine caused by the pressure drop of the engine exhaust at the evaporator shell side is far lower than the former. When the evaporation pressure is kept constant, the power output of the diesel engine–ORC combined system gradually increases with the increase of engine speed. The main reason for this condition is that with the increase of engine speed, on one hand, the available exhaust energy from the diesel engine increases, then the net power output of the ORC system increases. On the other hand, the power loss of diesel engine caused by the pressure drop of engine exhaust at the evaporator shell side also increases, which is far lower than the former. The maximum value of the power increase rate of the diesel engine–ORC combined system is 6.5%.

As shown in Figure 19, when the engine speed is kept constant, the BSFC decrease rate of the diesel engine–ORC combined system gradually increases with the increase of evaporation pressure. When the evaporation pressure is kept constant, the BSFC decrease rate of the diesel engine–ORC combined system gradually increases with the increase of engine speed. The maximum value of the BSFC decrease rate of the diesel engine–ORC combined system is 6.1%.

5. CONCLUSIONS

In this paper, a CFD model of a fin-and-tube evaporator of an ORC system is established by using Fluent software. The flow and heat transfer characteristics of the engine exhaust at the evaporator shell side are obtained. The performance of the fin-and-tube evaporator is analyzed based on the field synergy principle. A simulation model of the diesel engine is built based on GT–Power under various operating conditions. The influence of the pressure drop of the engine exhaust at the evaporator shell side on diesel engine performance is studied. The variation tendency of the power output and BSFC of diesel engine–ORC combined system are discussed.

(1) The exhaust gas temperatures decrease gradually along the flow direction at the evaporator shell side. The high-temperature exhaust gas in the flow space between the first and second rows diffuses quickly. To ensure that the exhaust flow is evenly distributed in the spaces between the neighboring fin layers, a reasonable design of the shapes that connect the main body with the front end and back end portions is necessary.

(2) The field synergy effect for the areas among the tube bundles of the evaporator main body and the field synergy effect for the areas among the fins on the windward side are satisfactory. However, the field synergy effect for the areas among the fins on the leeward side is weak. To further improve the heat transfer performance, the layout of the fins can be modified, and the shapes and angles of the fins can be adjusted to make the synergy angle between the velocity and the temperature gradient close to 0° or 180°, which can provide a theoretical guide for the design and performance optimization of the fin-and-tube evaporator.

(3) As a result of the pressure drop caused by the evaporator of the ORC system, the diesel engine power and torque decrease slightly, whereas the BSFC increases slightly with the increase of the exhaust back pressure. With the increase of engine speed, the power loss, the torque

loss, and the BSFC increment increase gradually, with the most significant change being less than 1%.

(4) Compared with the diesel engine itself, the maximum increase of power output is 6.5% for the diesel engine–ORC combined system, and the maximum decrease of BSFC is 6.1%.

In this research, we have done some experimental and simulation work for the existing evaporator of ORC system in our laboratory. In the near future, the heat transfer performance of the evaporator and the influence of pressure drop on the operating characteristics of the diesel engine–ORC combined system will be discussed for different kinds of evaporators. To further improve the heat transfer performance of the evaporator and the operating performance of the diesel engine–ORC combined system, the flow and heat transfer characteristics of the working fluid at the evaporator tube side should be investigated.

ACKNOWLEDGMENTS

This work was sponsored by the National Natural Science Foundation of China (Grant No. 51376011), the Beijing Natural Science Foundation Program (Grant No. 3152005), the Scientific Research Key Program of Beijing Municipal Commission of Education (Grant No. KZ201410005003), the National Basic Research Program of China (973 Program) (Grant No. 2011CB707202), and the Key Project of Thirteenth Scientific Research Foundation for Graduate Students in Beijing University of Technology (Grant No. ykj-2014-10708).

AUTHOR CONTRIBUTIONS

Chen Bei wrote the main body of the paper. Hongguang Zhang and Fubin Yang revised the paper. Songsong Song, Enhua Wang, and Hao Liu performed the simulations. Ying Chang, Hongjin Wang, and Kai Yang performed the experiments. All authors read and approved the manuscript.

CONFLICTS OF INTEREST

The authors declare no conflict of interests.

REFERENCES

1. Dolz, V.; Novella, R.; García, A.; Sánchez, J. HD Diesel Engine Equipped with a bottoming Rankine Cycle as a Waste Heat Recovery System Part I: Study and Analysis of the Waste Heat Energy. Appl. Therm. Eng. **2012**, 36, 269–278.
2. Heywood, J.B. Internal Combustion Engine Fundamentals; McGraw-Hill, Inc.: New York, NY, USA, 1988.
3. Kaska, O. Energy and exergy analysis of an organic Rankine for power generation from waste heat recovery in steel industry. Energy Convers. Manag. **2014**, 77, 108–117.
4. Karellas, S.; Leontaritis, A.D.; Panousis, G.; Bellos, E.; Kakaras, E. Energetic and exergetic analysis of waste heat recovery systems in the cement industry. Energy **2013**, 58, 147–156.
5. Wang, J.F.; Yan, Z.Q.; Zhao, P.; Dai, Y.P. Off-design performance analysis of a solar-powered organic Rankine cycle. Energy Convers. Manag. **2014**, 80, 150–157.
6. Uris, M.; Linares, J.I.; Arenas, E. Techno-economic feasibility assessment of a biomass cogeneration plant based on an Organic Rankine Cycle. Renew. Energy **2014**, 66, 707–713.
7. Uusitalo, A.; Honkatukia, J.; Turunen-Saaresti, T.; Larjola, J. A thermodynamic analysis of waste heat recovery from reciprocating engine power plants by means of organic Rankine cycles. Appl. Therm. Eng. **2014**, 70, 33–41.
8. Leslie, N.P.; Zimron, O.; Sweetser, R.S.; Stovall, T.K. Recovered energy generation using an organic Rankine cycle system. ASHRAE Trans. **2009**, 115, 220–230.
9. Imran, M.; Usman, M.; Park, B.S.; Kim, H.J.; Lee, D.H. Multi-objective optimization of evaporator of organic Rankine cycle (ORC) for low temperature geothermal heat source. Appl. Therm. Eng. **2015**, 80, 1–9.
10. Zhang, H.G.; Wang, E.H.; Fan, B.Y. Heat transfer analysis of a finned-tube evaporator for engine exhaust heat recovery. Energy Convers. Manag. **2013**, 65, 438–447.
11. Kundu, B.; Das, P.K. Performance and optimum dimensions of flat fins for tube-and-fin heat exchangers: A generalized analysis. Int. J. Heat Fluid Flow **2009**, 30, 658–668.
12. Zhang, Z.G.; Ma, D.B.; Fang, X.M. Experimental and numerical heat transfer in a helically baffled heat exchanger combined with one three-dimensional finned tube. Chem. Eng. Process. **2008**, 47, 1738–1743.

13. Ozden, E.; Tari, I. Shell side CFD analysis of a small shell-and-tube heat exchanger. Energy Convers. Manag. **2010**, 51, 1004–1014.]
14. Zhang, J.F.; Li, B.; Huang, W.J.; Lei, Y.G.; He, Y.L.; Tao, W.Q. Experimental performance comparison of shell-side heat transfer for shell-and-tube heat exchangers with middle-overlapped helical baffles and segmental baffles. Chem. Eng. Sci. **2009**, 64, 1643–1653.
15. An, C.S.; Choi, D.H. Analysis of heat-transfer performance of cross-flow fin-tube heat exchangers under dry and wet conditions. Int. J. Heat Mass Transf. **2012**, 55, 1496–1504.
16. Bhuiyan, A.A.; Amin, M.R.; Sadrul Islam, A.K.M. Three-dimensional performance analysis of plain fin tube heat exchangers in transitional regime. Appl. Therm. Eng. **2013**, 50, 445–454.
17. Tang, L.H.; Zeng, M.; Wang, Q.W. Experimental and numerical investigation on air-side performance of fin-and-tube heat exchangers with various fin patterns. Exp. Therm. Fluid Sci. **2009**, 33, 818–827.
18. Aytunc, E.; Bails, O.; Levent, B. Effect of geometrical parameters on heat transfer and pressure drop characteristics of plate fin and tube heat exchangers. Appl. Therm. Eng. **2005**, 25, 2421–2431.
19. Şahin, H.M.; Dal, A.R.; Baysal, E. 3-D Numerical study on the correlation between variable inclined fin angles and thermal behavior in plate fin-tube heat exchanger. Appl. Therm. Eng. **2007**, 27, 1806–1816.
20. Han, H.; He, Y.L.; Li, Y.S.; Wang, Y.; Wu, M. A numerical study on compact enhanced fin-and-tube heat exchangers with oval and circular tube configurations. Int. J. Heat Mass Transf. **2013**, 65, 686–695.
21. Tao, Y.B.; He, Y.L.; Huang, J.; Wu, Z.G.; Tao, W.Q. Numerical study of local heat transfer coefficient and fin efficiency of wavy fin-and-tube heat exchangers. Int. J. Therm. Sci. **2007**, 46, 768–778.
22. Hatami, M.; Ganji, D.D.; Gorji-Bandpy, M. Numerical study of finned type heat exchangers for ICEs exhaust waste heat recovery. Case Stud. Therm. Eng. **2014**, 4, 53–64.
23. Bai, S.Q.; Lu, H.L.; Wu, T.; Yin, X.L.; Shi, X.; Chen, L.D. Numerical and experimental analysis for exhaust heat exchangers in automobile thermoelectric generators. Case Stud. Therm. Eng. **2014**, 4, 99–112.
24. Diesel fuel. Available online: http://en.wikipedia.org/wiki/Diesel_fuel (accessed on 14 March 2015).
25. Lemmon, E.W.; Huber, M.L.; McLinden, M.O. REFPROP Version 8.0, NIST Standard Reference Database 23; US Department of Commerce: Washington, DC, USA, 2007.

26. Cengel, Y.A.; Boles, M.A. Thermodynamics—An Engineering Approach, 6th ed.; McGraw-Hill: New York, NY, USA, 2008; pp. 423–469.
27. Poling, B.E.; Prausnitz, J.M.; O'Connell, J.P. The Properties of Gases and Liquids, 5th ed.; McGraw-Hill: New York, NY, USA, 2001.
28. Roache, P.J.; Ghia, K.N.; White, F.M. Editorial policy statement on the control of numerical accuracy. J. Fluids Eng. **1986**, 108.
29. Guo, Z.Y.; Li, D.Y.; Wang, B.X. A novel concept for convective heat transfer enhancement. Int. J. Heat Mass Transf. **1998**, 41, 2221–2225. [
30. Tao, W.Q.; Guo, Z.Y.; Wang, B.X. Field synergy principle for enhancing convective heat transfer—its extension and numerical verification. Int. J. Heat Mass Transf. **2002**, 45, 3849–3856.
31. Guo, Z.Y.; Tao, W.Q.; Shah, R.K. The field synergy (coordination) principle and its applications in enhancing single phase convective heat transfer. Int. J. Heat Mass Transf. **2005**, 48, 1797–1807.
32. Wang, E.H.; Zhang, H.G.; Fan, B.Y.; Ouyang, M.G.; Yang, K.; Yang, F.Y.; Li, X.J.; Wang, Z. 3D numerical analysis of exhaust flow inside a fin-and-tube evaporator used in engine waste heat recovery. Energy **2015**, 82, 800–812.

10

CHAPTER

3D VISUALIZATION OF THERMAL RESISTANCE AND CONDENSATION PROBLEMS USING INFRARED THERMOGRAPHY FOR BUILDING ENERGY DIAGNOSTICS

Youngjib Ham[1*] and Mani Golparvar-Fard[2]

[1]*Department of Civil and Environmental Engineering, University of Illinois at Urbana-Champaign, Urbana, IL, USA.*

[2]*Department of Civil and Environmental Engineering, Department of Computer Science, University of Illinois at Urbana-Champaign, Urbana, IL, USA.*

ABSTRACT

Background

Building deteriorations instigated by material degradations or moisture intrusions are the primary causes for energy inefficiency in many existing buildings. For choosing appropriate retrofits, it is important to carefully diagnose and analyze building areas in need of improvements. In addition to reliable sensing and analysis of as-is energy performance, an intuitive recording and visualization of energy diagnostic outcomes are also critical to effectively illustrate the as-is building conditions to homeowners during retrofit decision-making processes.

Method

Toward this goal, this paper presents a thermography-based method to visualize the actual thermal resistance and condensation problems in 3D while taking static occlusions into account. First, several overlapping digital and thermal images are collected from the building areas under inspection. Using a computer vision method– consisting of image-based 3D point cloud and mesh modeling algorithmsactual 3D spatio-thermal actual 3D spatio-thermal models are generated where surface temperature can be queried at the level of 3D points. Based on the resulting 3D spatio-thermal models and by measuring the reflected and dew point temperatures, the actual R-values of building assemblies are calculated, and the condensation issues are analyzed. Taking static occlusions into account, (1) the distribution of the actual thermal resistance over each building assembly, (2) the detected building areas with condensation problems, and (3) the corresponding geometrical and thermal characteristics are jointly visualized within a 3D environment.

Results

To validate the method and investigate the perceived benefits, four experiments have been conducted in existing buildings. Surveys are also conducted by professional energy auditors. The proposed method provides 3D visual representation of the actual thermal resistance distributions and building areas associated with condensation issues at the level of 3D points across geometrical forms while taking static occlusions into account.

Conclusions

The experimental results and the feedback received from the professionals show the promise of the proposed method in facilitating systematic post-examination of building deteriorations and support retrofit decision-makings. Ultimately, converting surface temperature data obtained from an IR camera into 3D visualization of energy performance metrics and possible condensation problems enables practitioners to better understand the as-is building conditions.

Keywords: Image-based 3D Reconstruction Building condition assessment Building condensation problem Thermal resistance Building energy diagnostics

INTRODUCTION

The emerging energy crisis of the building sector and the legislative measures on improving energy efficiency are steering the construction industry towards adopting new design concepts and construction methods that can decrease the overall energy loads. For example, in order to lower the energy cost required for space conditioning, building practitioners are embracing the concept of net-zero passive houses at the design phase. The use of building materials with higher thermal resistance is recommended for better insulation at the construction phase.

The problem of energy inefficiency is however not limited to the design and construction of new buildings. According to a recent report by the U.S. Department of Energy (U.S. DOE [2010]), around 35% of input energy in existing buildings in the U.S. is still being wasted. A primary source of these inefficiencies is thermal performance problems in the building envelope, which can cause an excess in energy consumption for space heating and cooling purposes. Roth et al. ([2005]) reported that such faulty behaviors can account for up to 10% of the total energy consumption in existing buildings during their operational phase. Despite their significance, identifying and assessing building performance problems and improving the energy efficiency under such conditions is not trivial.

Today, building practitioners widely use Infrared (IR) thermography to identify sources of energy loss in building envelopes. The IR thermography –which detects and measures heat variations from surfaces in a non-destructive manner– is particularly known for its usefulness for identifying thermal defects. Nonetheless, current thermographic inspection processes have several inefficiencies. For example, thermal images captured using consumer-level IR thermal cameras typically have low spatial resolutions (160 × 120 or 320 × 240 pixel) and small fields-of-view compared to their digital counterparts (>1 megapixel). As a result, sensing thermal performance data for buildings, even a single room, requires large numbers of thermal images (e.g., ~300 to 400 thermal images for around $4m \times 5m$ room based on experiments). Considering significantly large collections of 2D thermal images that need to be manually analyzed for the purpose of assessing thermal conditions, their direct application for practical building diagnostics can be time-consuming and labor-intensive. In order to improve the process of sensing thermal performance using raw 2D thermal images, several studies (Borrmann et al. [2013]; [Ham and Golparvar-Fard 2013]; Lagüela et al. [2013]; Wang et al. [2013]) have recently focused on developing methods for 3D thermal modeling of building environments by leveraging digital photogrammetry or laser scanning, and promising results have been reported.

These methods for 3D thermal modeling provide a detailed representation of the actual temperature deviations on building surfaces, which is very effective for energy diagnostics. Nevertheless, the problem of characterizing and visualizing performance metrics and thermal problems from these models for post-inspection is still unexplored. The analysis of abnormal thermal regions – either from 2D thermal images or 3D thermal models – is still primarily based on the interpretation of the measured surface temperatures such as hot or cold spots. Yet, the surface temperatures alone do not directly represent the energy performance problems. Due to a lack of a benchmark, the interpretation of energy performance problems is still primarily based on the auditors' experiences and knowledge, which can cause the outcomes to be subjective or inconsistent. Considering a limited number of well-trained professionals, it is also likely that auditors' knowledge and experience would adversely impact the quality of the inspection.

Another important issue – which is often not considered– is that the deterioration rate of building assemblies typically vary even over small surface areas. For instance, when it comes to building envelopes, various regions across the surface may suffer from different levels of degradation. As such, their conditions would change differently over time. Therefore, a method for building diagnostics should assess and visualize the as-is building conditions at point-level across geometrical forms of the building elements so that defects can be precisely detected and localized.

The presence of static occlusions in existing buildings has also raised concerns about the applicability of thermography-based methods for building diagnostics. Ideally, for accurate measurement of surface thermal performance, any objects around the building interior surfaces need to be removed prior to thermographic inspections. However, in practice, for most in-use buildings, it is not trivial to take away all objects that block the line of sight due to the immobility of objects or space limitations. Theoretically, there is no way to measure the thermal performance for the building areas that are occluded by objects, except for encouraging practitioners to remove as many objects as possible before thermographic inspection. Thus, this issue remains as an open challenge.

To address these challenges, this paper presents a new interactive method for 3D visualization of the actual R-value of building assemblies and condensation problems. The proposed method builds upon the authors' prior computer vision research on 3D spatio-thermal modeling using large numbers of unordered thermal imagery collected by a consumer-level IR camera [(Ham and Golparvar-Fard 2013)]. Beyond synthesizing raw 2D surface temperature data, the method extracts and visualizes building energy performance indicators from the collected thermal imagery. It facilitates the process of identifying potential performance problems by allowing auditors to spend less time on analyzing large amounts of thermal images, and instead focus on investigating the causes of the problems and analyzing various retrofit alternatives.

The primary technical contributions of this paper are two-fold: (1) a method for 3D visualization of energy performance metrics and moisture issues at the level of 3D points across geometrical forms of building elements; and (2) a method for minimizing the adverse impact of static occlusions on thermographic inspection in real-world building environments. In the following sections, previous works on measuring R-values of building materials using thermal imagery and analyzing condensation problems are reviewed, and their challenges in applying to building condition assessment are discussed. Next, the research objective, underlying environmental assumptions, and the proposed visualization method are presented in detail. Finally, experimental results on four building assemblies in two case studies and the responses from domain expert evaluations are discussed.

BACKGROUNDS

Thermal resistance (R-value) of building assemblies

Analyzing heat transfer conditions through building envelopes is now a required step in rating the energy performance levels of existing buildings (Fokaides and Kalogirou [2011]). This is because during the operational phase, the thermal resistance of building materials gradually decreases. Thus, the actual R-value of the building assemblies with thermal defects is typically lower than the notional value declared by their manufactures. Declining R-values of the building materials results in unnecessary heat transfer through building façades. This in turn increases the operational frequency of heating and cooling systems for space conditioning purposes. Hence, beyond using building materials with high R-value during the design and construction, it is important to explore how the as-is thermal resistance of building assemblies lowers over time.

Recently, several research groups have focused on the problem of non-destructive measurement of the actual heat transfer conditions of building assemblies using thermal images. Madding's work (Madding [2008]) is one of the earliest studies that propose a method for analyzing thermal images to estimate the actual heat transfer conditions for building envelopes. Based on the environmental assumption of a steady-state heat transfer condition of building environments, this work measured the R-value of drywall assemblies using the indoor surface temperature data obtained from thermal images. The measurements were experimentally conducted in controlled lab environments, and then they were compared with the theoretical expected properties. Likewise, based on the similar environmental assumptions, methods proposed in (Albatici and Tonelli [2010]; Dall'O' et al. [2013]; Fokaides and Kalogirou [2011]) measured the overall heat transfer coefficient of the building assemblies. For validation, several experiments were conducted for a few standard wall assemblies. In these studies, the difference between the measured values using thermal images and the notional values was reported to be in the range of ~10% (Madding [2008]), 10-20% (Fokaides and Kalogirou [2011]), and around 15% for cavity walls (Dall'O' et al. [2013]).

For quantifying the as-is thermal resistance of building assemblies, the current non-destructive method regulated by the international standard is to position heat flux meters (HFMs) –a transducer that produce an electrical signal proportional to the heat rate– on the representative building areas and directly measure the heat flow (International Organization for Standardization (ISO) [2007]). However, considering possible variations of the thermal conditions even over small surface areas, it is difficult to ensure that the thermal resistance measured by using a few HFMs will actually represent the entirety of an existing building envelope (Albatici and Tonelli [2010]; Dall'O' et al. [2013]). Interestingly, studies such as (Albatici and Tonelli [2010]) and (Dall'O' et al. [2013]) report 40-80% and 40-175% error between the measurements of the overall heat transfer coefficient using HFMs and their theoretical expected values. Although these observations are primarily based on limited experiments, when one considers the significance of these errors, the applicability of thermal imagery for quantifying the actual heat transfer condition of building assemblies becomes more attractive.

Despite such promising benefits of using thermography, the direct application of 2D thermal imagery as used in prior works is still challenging. In prior studies, a single or at most a few temperature data were extracted from the designated areas captured in 2D thermal imagery, and then a single measurement of the actual heat transfer condition was performed for each building assembly. Hence, these methods assumed that the obtained thermal resistance is representative of all spots on the inspected building assemblies. However, because of different deterioration rates, the actual heat transfer conditions may vary over small scales across geometrical forms. Thus, single measurements will not represent the dynamic variations in the actual R-values that are typically caused by non-uniform deteriorations in the building materials. More importantly, interactive 3D visualization of these dynamic variations in the context of the building under inspection will provide an opportunity for practitioners to better understand the as-is heat transfer conditions.

Condensation problems in building environments

In building environments, water vapor condensation happens wherever exposed building surfaces are at a lower temperature than the dew point. Typically, it is not easy to detect small amounts of the condensed water on building surfaces with the naked eye. Failure in detecting the onset of such condensation issues causes the condensed moisture to accumulate over time. Then, molds and dust mites may build up, which ultimately has potential for causing allergic reactions among the building occupants. In addition, the resulting wet surface may cause flaking paints and peeling wallpapers. This is when the building occupants are more likely to detect the problems with their naked eyes. In building environments, windowsills or thermal bridges around corners of the walls typically have low temperatures. Thus, energy auditors need to proactively analyze these areas whether the surface temperature has reached to the dew point.

Several studies have focused on analyzing the condensation phenomenon on building façades and the risk associated with the occurrence. Bellia and Minichiello analyzed the thermal and moisture performance of building assemblies, and focused on surface condensation issues of building façades and interstitial condensations within multi-layer walls (Bellia and Minichiello [2003]). Aelenei and Henriques investigated the external environmental conditions that cause condensation problems in building environments (Aelenei and Henriques [2008]). They concluded that the thermal convection effect and the amount of moisture in the air are the most influential factors in forming condensation problems on building façades. These prior works primarily examined the sources that cause condensation and identified ways that can reduce the risk associated with these problems. Nonetheless, devising energy auditing methods that can easily detect condensation problems in building environments is still an open research area. Today, by comparing the surface and dew point temperature, thermal cameras enable 2D detection of those building areas that suffer from condensation issues (Hoff [2011]). These detections are typically done on a single image displayed on the screen of the thermal camera. Because these images are not geo-tagged, it is often not easy to figure out at a later stage which building areas are associated with the detected condensation problems. Moreover, these 2D assessments do not provide any information about the surface area of the detected condensation problems. Hence, it is difficult at a later stage to quantitatively estimate the significance of their impact.

OVERVIEW OF THE PROPOSED METHOD

This paper presents and validates a 3D visualization method for quantitative assessment of the as-is building conditions. The method leverages a paired collection of digital and thermal images taken using a single IR thermal camera from the building environments under inspection, and accounts for static occlusions in the building environments. The hypothesis is that the quality of the current building condition assessment procedures can be improved with intuitive 3D visualizations that enable practitioners to (1) explore the distribution of the actual heat transfer conditions for building assemblies; and (2) detect condensation problems in 3D. The overall workflow is illustrated in Figure 1.

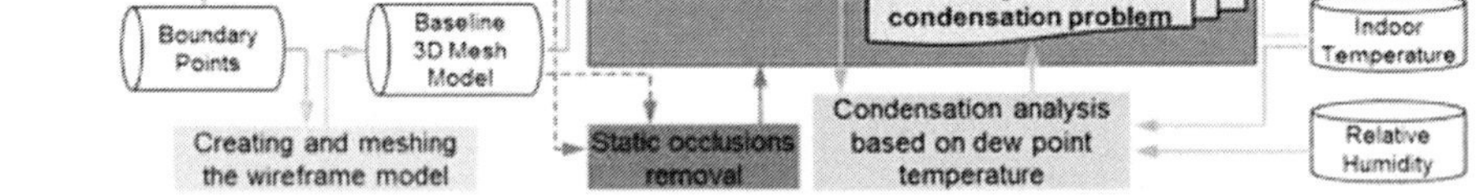

Figure 1. An overview of the data and process in the proposed 3d visualization method for building condition assessment using infrared thermography.

By using digital and thermal images, the proposed method first generates the 3D spatio-thermal mesh models which contain the surface temperature readings at the level of 3D vertices. Then, with environmental measurements, actual thermal resistances are calculated, and building surface areas suffering from condensation problems are detected at the level of 3D vertices in the mesh. Finally, taking possible static occlusions into account, the distribution of the actual R-value, the detected building areas with condensation problems, and the corresponding geometrical and thermal characteristics are interactively visualized within a common 3D environment. In the following sections, the underlying environmental assumptions, experimental setups, and each step of the proposed method are presented in detail.

ENVIRONMENTAL ASSUMPTIONS AND EXPERIMENTAL SETUPS

The proposed method is based on the following environmental assumptions: (1) the indoor building environment is assumed to have a quasi-steady-state condition of heat transfer during the thermographic inspection process. This is consistent with all recent works on non-destructive measurement of thermal properties (Albatici and Tonelli [2010]; Dall'O' et al. [2013]; Fokaides and Kalogirou [2011]; Ham and Golparvar-Fard [2014]; Madding [2008]). For achieving a quasi-steady-state condition of heat transfer, the building envelopes were not exposed to direct solar radiations and wind loadings as the inspections were conducted before sunrise on a non-windy and cloudy day. This minimizes the influence of the convective heat loss and the surface temperature rise on the exterior of building envelopes. As such, the possible increase in building surface temperatures caused by the release of the absorbed solar energy throughout a day was minimized; 2) it was further assumed that the heat transfer between the indoor surfaces and the thermal camera lens is due to thermal convection and radiation. This assumption is consistent with prior works (Fokaides and Kalogirou [2011]; Ham and Golparvar-Fard [2014]; Madding [2008]). In addition, the building interior spaces were heated for a few hours prior to thermographic inspections, which allowed to clearly capture the as-is heat transfer conditions through building façades during the winter season.

For testing the proposed 3D visualization and conducting professional surveys, several experiments were conducted on four wall assemblies and windows adjacent to the exterior in an existing residential building (Case #1) and an instructional facility (Case #2). For 3D spatio-thermal modeling, digital and thermal images were captured by using a FLIR E60 thermal camera that has a built-in digital camera. Each pair of digital and thermal images were captured simultaneously. The technical specification of the thermal camera used in experiments is presented in Table 1. When temperature deviations used for calculating thermal resistances (Measuring actual R-values at the level of 3D points) are too small, accurate measurement using inexpensive consumer-level IR cameras would be challenging. To alleviate this, similar to (Madding [2008]), the impact of the systematic error on the measurement accuracy was minimized by using a single device for measuring temperature differences.

Table 1 Technical specification of the thermal camera used for data collection

Camera technical items	Values
Built-in digital camera resolution	2048 × 1536 pixels
IR resolution	320 × 240 pixels
Thermal sensitivity	< 0.05°C
NETD	50 mK
Measurement accuracy	2°C or ±2% of reading

METHOD

Reconstructing 3D building thermal performance using 2D thermography

The first step to assess and visualize as-is building conditions at the level of 3D points is to model the actual thermal performance of the building under inspection in 3D. The method for generating 3D spatio-thermal models of building environments builds upon the authors' recently prototyped methods [(Ham and Golparvar-Fard 2013)] and (Golparvar-Fard and Ham [2014]). First, both thermal and digital imagery are collected by practitioners using a consumer-level IR camera. Then, the 3D geometrical point cloud of the building is automatically generated by using the collected digital images. Here, an image-based 3D reconstruction method is used, which consists of Graphic Processing Unit (GPU)-based Structure-from-Motion (SfM) and Multi-View Stereo (MVS) algorithms. The next step is 3D thermal modeling using the collected thermal images. Unlike digital images, thermal images are typically featureless in computer vision terms and have low spatial resolutions. The lack of distinct visual features impede the application of the same image-based 3D modeling algorithm used on digital images for 3D building geometrical modeling purposes. Instead, through a one-time process, the thermal camera is internally calibrated, and the relative pose of the thermal camera lens with respect to the digital camera lens on the IR thermal camera is estimated. Based on the obtained intrinsic and extrinsic thermal camera parameters, a 3D thermal point cloud model is generated using the dense reconstruction algorithm. The resulting 3D thermal and geometrical point cloud models are automatically superimposed, forming *3D spatio-thermal* point cloud models. Next, the outcome is converted to a 3D thermal mesh model by using *k-d* tree structure and the nearest neighborhood searching algorithm. In this process, a geometrical baseline mesh is formed by using the boundary points extracted from the building geometrical point cloud as the basis for the transformation. The nearest 3D thermal point with respect to each vertex is searched for in the baseline mesh, and then the associated thermal values are mapped to the vertices. Figure 2 illustrates 3D thermal mesh modeling of two different building environments (top and bottom: Case #1 and #2 respectively).

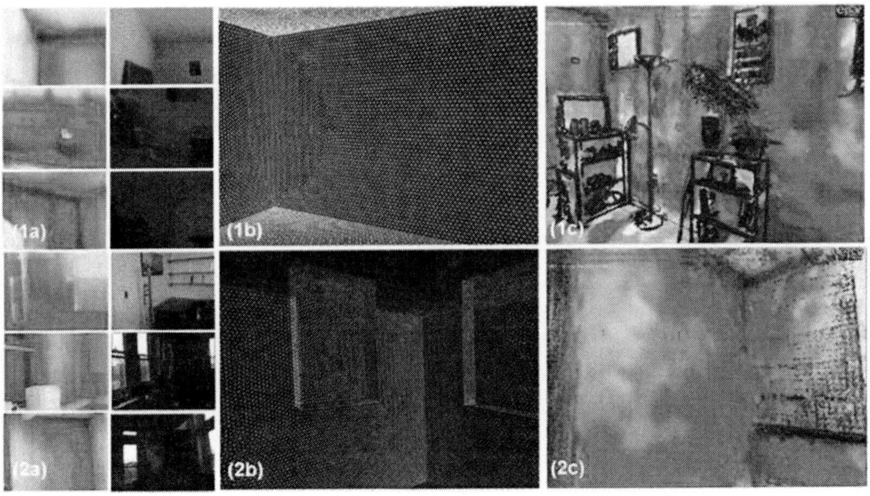

Figure 2. 3D THERMAL MESH MODELING. (A): Unordered digital and thermal imagery, **(B)**: Geometrical baseline mesh models, **(C)**: 3D thermal mesh models of interior surfaces jointly represented with building geometrical point cloud.

Here, different from previous works [(Ham and Golparvar-Fard 2013)] and (Golparvar-Fard and Ham [2014]), and to better deal with static occlusions in building environments, a threshold is used to identify those thermal points that have a certain distance from the geometrical baseline mesh. It is noted that the thermal points may belong to objects on or around the wall, not the building assembly itself. Thus, when searching for the nearest thermal points to each vertex in the baseline mesh, a geometrical threshold based on the registration error of the point cloud and the baseline mesh is applied. At a later stage, the points from non-relevant objects are discarded and are not used for calculations. The details will be discussed in Section (Considering static occlusions in 3D visualization of the as-is building conditions).

Measuring actual R-values at the level of 3D points

The environmental measurements in this paper build on the environmental assumptions presented in the section (Environmental Assumptions and Experimental Setups), which is consistent with recent works on non-destructive measurement of R (or U)-value using thermography (Albatici and Tonelli [2010]; Dall'O' et al. [2013]; Fokaides and Kalogirou [2011]; Ham and Golparvar-Fard [2014]; Madding [2008]). Under such environmental assumptions, the overall heat transfer rate (dQ/dt) through a building surface with the area of (A), a thermal resistance (R), and a temperature difference of ΔT between inside and outside can be described using the following equation:

$$\frac{dQ}{dt} = \frac{1}{R} \times A \times \Delta T \tag{1}$$

Here, the overall heat transfer (Q) in building environments can be described by the combination of thermal convection and radiation. First, the heat transfer caused by thermal convection between the interior surface of building environments and the indoor air is calculated using the following equation:

$$Q_{Convective} = \alpha_{convective} \times A \times \left| T_{inside,\ air} - T_{inside,wall} \right| \tag{2}$$

Where $\alpha_{convective}$ is the convective heat transfer coefficient which is a function of the airflow (e.g., laminar or turbulent) and the temperature deviations between building surfaces and indoor air. Typical convective heat transfer coefficients of indoor building environments are within the range of 2–8 W/m^2K. In this work, this value is adopted based on (Dall'O' et al. [2013]). As part of data collection, the inside and outside air temperatures, $T_{inside,\ air}$ and $T_{outside,\ air}$, are measured using a thermometer. Next, the heat transfer between the interior building surface and the surroundings caused by thermal radiation is calculated using the following equation:

$$Q_{Radiation} = \varepsilon \times \sigma \times A \times \left| T^4_{inside,\ wall} - T^4_{inside,\ reflected} \right| \quad (3)$$

Where ε is thermal emissivity and σ is Stefan-Boltzmann constant ($5.67 \times 10^{-8}\ W/m^2K^4$. For thermal emissivity, the measurement in (Mikron Instrument Company [2012]; Singham [1962]) was adopted. To measure the reflected temperature ($T_{inside,\ reflected}$), a small crumpled aluminum foil was placed on the inspection areas prior to the thermographic inspection (Figure 3). Since the crumpled aluminum foil has low emissivity and high reflectivity, the reflected temperature can be robustly measured from the crumpled foil (FLIR system [2010]). Prior works (Fokaides and Kalogirou [2011]; Madding [2008]) experimentally demonstrated that measuring the difference between surface and reflected apparent temperatures is one of the most sensitive factors on reliable R-value measurement in a non-destructive form. Here, instead of extracting a single temperature from the crumpled foil or averaging temperatures of a few target points as used in (Fokaides and Kalogirou [2011]; Madding [2008]), the average temperature is calculated from all pixels that represent the crumpled foil. To do so, the RGB color values of these pixels from the thermal image (Figure 3c) were converted into the corresponding absolute temperature values based on a normalized temperature spectrum.

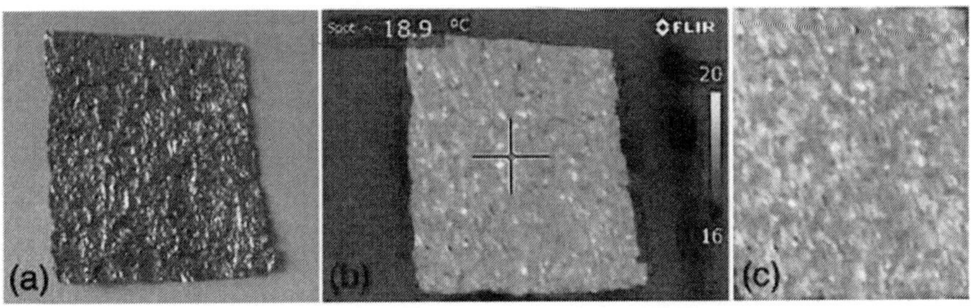

Figure 3. MEASURING REFLECTED TEMPERATURE. (A): crumpled aluminum foil located on the inspection surface, **(B)**: thermal image simultaneously captured from the same viewpoint, **(C)**: temperature readings on the crumpled aluminum foil surface.

It is assumed that interior surfaces in the given space under inspection have the same reflective temperatures, which is experimentally validated in (Fokaides and Kalogirou [2011]). Finally, by combining the Equations (1), (2), and (3), the thermal resistance can be described by using the following equation:

$$R = \frac{\left|T_{inside,\ air} - T_{outside,\ air}\right|}{\alpha_{con} \times \left|T_{inside,\ air} - T_{inside,wall}\right| + \varepsilon \times \sigma \times \left|T^4_{inside,\ wall} - T^4_{inside,\ reflected}\right|} \quad (4)$$

Within the resulting 3D spatio-thermal mesh models, $T_{inside,\ wall}$ can be queried from each 3D vertex. Here, different from prior works, this study focuses on measuring actual thermal resistance of building assemblies while taking possible static occlusions into account as discussed in the previous section.

Detecting building condensation problems in 3D

Building condensation is due to the fact that a given volume of air at a given temperature can only accommodate a certain amount of water. This is typically characterized by the relative humidity and the maximum absolute humidity. The absolute humidity is a measure of the amount of the water content in the air, which itself is a function of the air temperature. The relative humidity is a measure of the percentage of the actual water content in the air with respect to the maximum absolute humidity (i.e., the amount of maximum possible water content at a given air temperature). When the air contains more water content than the maximum possible amount at a given air temperature, the amount of the extra water vapor will condense on the surrounding surfaces. Hence, the dew point temperature (T_D) is a function of the relative humidity (f) and indoor temperature ($T_{inside,\ air}$) that can be described by the following equation (Wanielista et al. [1997]):

$$T_D = \left(\frac{f}{100}\right)^{1/8} \left(112 + 0.9T_{inside,air}\right) + 0.1T_{inside,air} - 112 \quad (5)$$

The indoor temperature as well as the relative humidity are measured with the use of a thermo-hygrometer as part of the thermographic inspection. In this work, the building surface temperature ($T_{inside,\ wall}$) can be queried at the level of 3D vertices from the generated 3D spatio-thermal mesh models. Thus, the building surface areas that are likely to suffer from condensation problems are directly detected by comparing $T_{inside,\ wall}$ and T_D at the level of 3D points. Here, the building surface areas relevant to condensation issues can be calculated from the mesh model. For each triangle face consisting of the three thermal points that their thermal values are below T_D, the 3D coordinates of the vertices are queried, and then the area is calculated by using the cross product of two corresponding vectors that form the triangle face.

Considering static occlusions in 3D visualization of the as-is building conditions

For minimizing the impact of static occlusions when visualizing performance metrics and thermal problems obtained from the analysis of the 3D thermal model, non-relevant objects is segmented from the reconstructed 3D scene. The idea is that if the nearest thermal point in the 3D spatio-thermal model is too far from the vertices in the geometrical baseline mesh, it is most likely that the thermal points belong to objects around the wall but not the building geometry itself. Hence, a threshold is set based on the point cloud-to-mesh registration error. To do that, the registration accuracy of the 3D spatio-thermal point cloud with the geometrical baseline mesh in the site coordinate system is examined. Mapping thermal point clouds to a geometrical baseline mesh involves transforming the 3D spatio-thermal model into the real-world site coordinate system. To determine this transformation, the closed-form solution using unit quaternions is used (Horn [1987]). The transformation from the 3D spatio-thermal model into the real-world has 7 degrees of freedom (1 for scaling +3 for translation +3 for rotation), and thus requires a minimum of three matching control points between the 3D spatio-thermal model and the real-world coordinate system Ham and Golparvar-Fard ([2013a]). These control points, *PiThermal*

and *PiSite*, can be extracted from the known dimensions in the building environments. This is solved by minimizing the sum of squared residual errors of the Euclidean transformation for (n) control points using the following:

$$\sum_{1}^{n} \| e^{i} \|^{2} = \sum_{1}^{n} \| p^{i}_{Site} - sR\left(P^{i}_{Thermal}\right) - T \|^{2} \qquad (6)$$

Where, *s*, *T*, and *R*(*PiThermal*) indicate the scaling factor, translational offset, and rotation of 3D spatio-thermal models respectively. The resulting registration error (e) is then used as a threshold for detecting thermal points that are likely to represent static occlusions. Among all $T_{inside,\ wall}$ that can be queried from each 3D vertices within the 3D spatio-thermal mesh, those thermal points belonging to non-relevant objects (i.e., above the threshold (e)) are removed as outliers and are not considered in the calculation of the thermal resistance of building assemblies and detecting condensation problems. Once the relevant-to-building geometry thermal points are queried from 3D spatio-thermal mesh models, the actual R-values of the building assemblies under inspection are calculated by using Equation (4), and the building areas associated with condensation issues are detected by using Equation (5).

Then, for 3D visualization, the measured R-values and the comparison of $T_{inside,\ wall}$ and T_D at the level of 3D points are converted into the corresponding RGB color value. To visualize the measured metrics and problems, the visualization platform provides several navigation modes such as a 3D free-flight navigation to efficiently visualize performance metrics and potential issues with the related building views. Similar to most engineering 3D viewers, when a camera is visited in the reconstructed 3D scene, the user can pan, tilt, and zoom-in & out for exploring the particular building areas in a mesh. For integrated visualization of the distribution of the actual R-value, the detected building areas with condensation issues, and the related geometrical and thermal characteristics, the proposed visualization platform leverages the following data structures:

1. $P_1^e, P_2^e, \ldots, P_n^e$: a set of relevant-to-building geometry 3D mesh vertices where $P_i^e = X_i^e, Y_i^e, Z_i^e, RGB_i^{R-value}, RGB_i^{Condensation}, RGB_i^{SurfaceTemperature}$ encapsulates the 3D location (X_i^e, Y_i^e, Z_i^e) the color value $(RGB_i^{R-value})$ of each vertex which corresponds to the measured R-value according to the given color spectrum, the color value $(RGB_i^{Condensation})$ of each vertex which is a metaphor based on traffic light colors to represent problematic and non-problematic areas using red and green colors respectively, and finally the color value $(RGB_i^{SurfaceTemperature})$ of each vertex associated with a temperature reading averaged from all thermal images that have observed the vertex
2. $P_1^g, P_2^g, \ldots, P_m^g$: a set of 3D points where $P_i^g = X_i^g, Y_i^g, Z_i^g, RGB_i^{Geometry}$ encapsulates the 3D location (X_i^g, Y_i^g, Z_i^g) and the color $(RGB_i^{Geometry})$ of each point in 3D building geometrical point cloud which is the average from the colors in all digital images that observed the point
3. $C_1^T, C_2^T, \ldots, C_l^T$: a set of thermal cameras representing the thermal camera projection matrix based on the relative pose with respect to the built-in digital camera
4. $C_1^D, C_2^D, \ldots, C_l^D$: a set of built-in digital cameras representing the digital camera projection matrix for each camera registered in a reconstructed 3D scene

5. A set of vertices that form each face
6. The number of vertices and faces in a mesh
7. The number of feature points and cameras that observe the points
8. Mapping relationship between points and cameras that observe the points

RESULTS AND DISCUSSIONS

The first five rows in Table 2 present the 3D thermal modeling metrics and the results from the case studies. In both cases, the computational time is benchmarked on a regular engineering workstation (an Intel i7 960 core with 24GBs RAM and NVIDIA GeForce GTX 400), and reasonably detailed 3D mesh models are generated in each case. Table 2 also shows the environmental measurements that were conducted during the experiments. Here, rather than concluding a single value of the actual thermal resistance for a building assembly in a deterministic way, Table 2 presents the range that the R-value for inspected building assemblies is most likely to be included.

Table 2 Experimental results of 3d thermal modeling and environmental measurements

Item	Metrics	Case #1	Case #2
3D thermal modeling	# of 2D Thermal Images	446	429
	# of 3D Thermal Points	3,042,350	2,838,478
	# of Vertices in Baseline Mesh Model (wall)	18,964	13,771
	# of Vertices excluding static occlusions in Mesh Model by using Threshold (wall)	17,297	7,740
	Computational Time for 3D Thermal Modeling	2hr 30min	2hr 21min
Environmental measurements	Outside air temperature (°C)	5	3
	Indoor air temperature (°C)	28	24
	Dew point temperature (°C)	13.15	11.64
	Relative humidity (%)	40	46
Calculated metrics in 3D	Error on registration (mm)	8.69	9.30
	Average (μ) of R-values of Wall (m^2K/W)	0.753	0.833
	Standard Deviation (σ) of R-values of Wall	0.10207	0.14463
	Range of actual R-values by considering 95% confidence ($\mu-1.96\sigma/\sqrt{n}$, $\mu+1.96\sigma/\sqrt{n}$)	(0.75168, 0.75472)	(0.82986, 0.83631)
	Average (μ) of R-values of Window (m^2K/W)	0.260	0.274
	Standard Deviation (σ) of R-values of Window	0.00173	0.00211
	Range of actual R-values by considering 95% confidence ($\mu-1.96\sigma/\sqrt{n}$, $\mu+1.96\sigma/\sqrt{n}$)	(0.26011, 0.26027)	(0.2737, 0.27389)

Figures 4, 5, 6, and 7 illustrate the distribution of the surface temperatures and the measured R-values at 3D points for wall assemblies and windows in case #1 and #2 respectively. Compared to prior works (Albatici and Tonelli [2010]; Dall'O' et al. [2013]; Fokaides and Kalogirou [2011]; Madding [2008]) which leverage a single measurement of actual thermal resistance for each building assembly using a single or a few temperature data, the proposed method can provide more detailed representation of the heat transfer conditions for building assemblies at the level of 3D points. Figures 8 and 9 show examples of visualizing the distribution of the R-values along with the corresponding building geometry in 3D. The 3D visualization of thermal resistances using different color gradients can provide an opportunity for building practitioners to easily and quickly understand the distribution of the actual thermal resistance in building environments. Because static occlusions are taken into account, the proposed method has potential for improving the accuracy of measurements. In both cases, by using the equation (6), the registration errors between 3D thermal point cloud and the geometrical baseline mesh are within a reasonable practical range (Table 2). Based on the calculated registration errors, it is observed that 8.8% (case #1) and 43.8% (case #2) of the thermal points were related to non-relevant objects around wall (e.g., Figures 8 and 9(b) and (c)). Thus, those thermal points are not considered when calculating the thermal resistances in 3D. Although it is attempted to minimize the impact of such static occlusions, false negatives are still possible due to incomplete segmentation. Examples are shown as oval-shaped areas in the Figures 8 and 9 (e), which causes inaccurate thermographic inspection for those areas. Here, the proposed method can also be used for the purpose of commissioning new buildings in addition to retrofit purposes during the operational phase. Unlike occupied buildings, there is minimal occlusion during commissioning inspection. Here, the proposed method can inspect if the new building is operating as expected by the design.

Beyond providing surface temperature data, the proposed method can also improve the quality of interpreting building thermography by detecting and localizing moisture problems on building envelopes and visualizing the outcome in 3D along with the corresponding geometrical characteristics. Figures 4, 5, 6, and 7 illustrate the dew point temperature (dotted line) and the distribution of surface temperatures for both case studies. These figures demonstrate noticeable deviations between these two temperatures and further highlight the prominence of the point-based measurements. Compared to the built-in option available in current thermal cameras (Hoff [2011]) for detecting condensation problems, the proposed method has

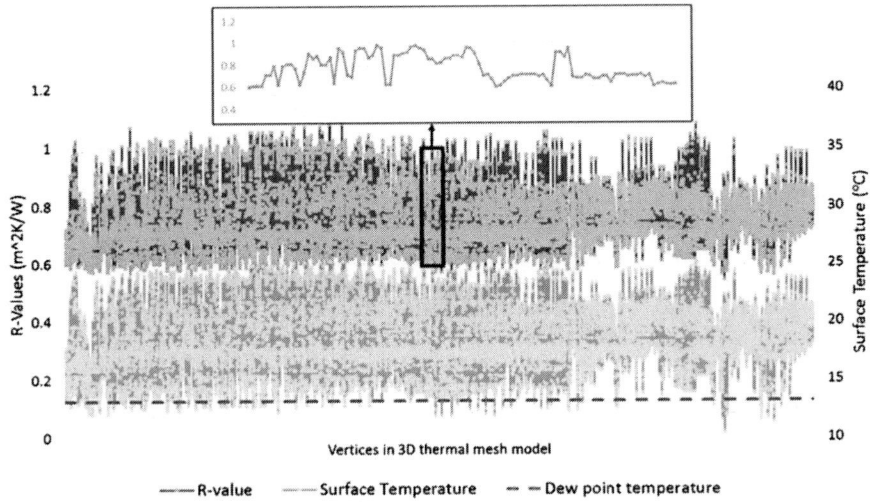

Figure 4 Point-level temperature and actual r-values for the wood frame walls in case study #1.

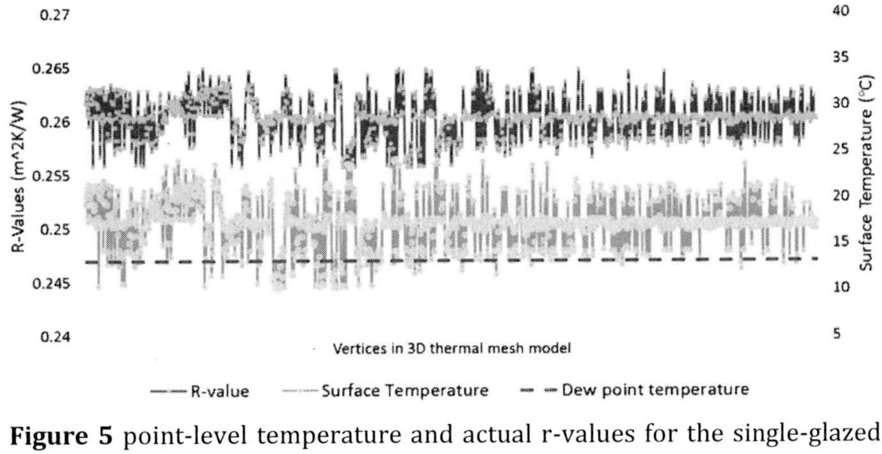

Figure 5 point-level temperature and actual r-values for the single-glazed clear window in case study #1.

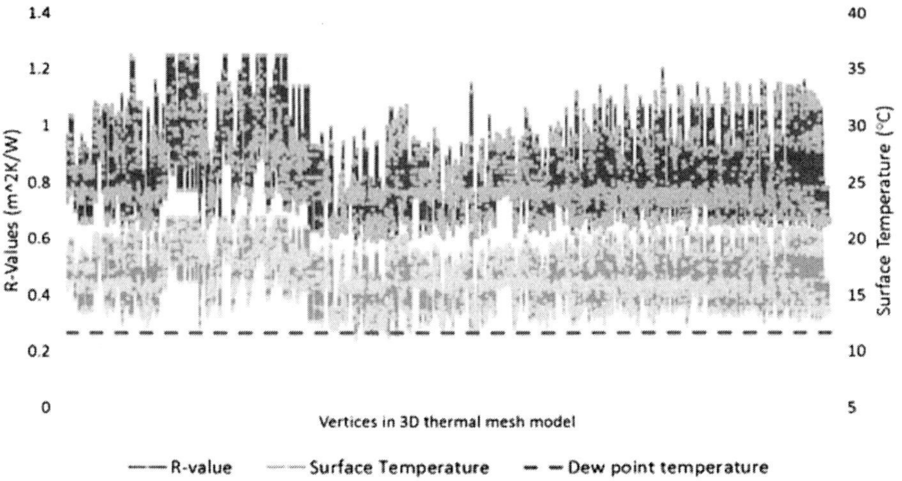

Figure 6. Point-level temperature and the actual r-values for concrete masonry walls in case study #2.

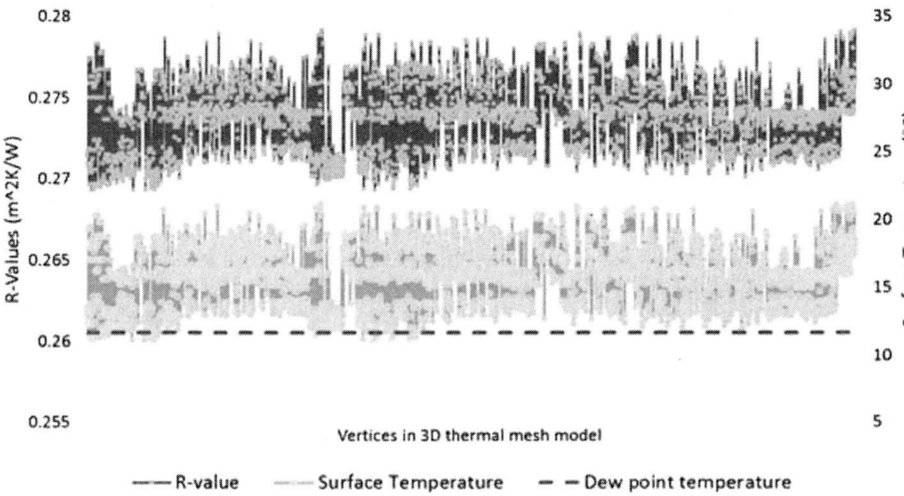

Figure 7 . Point-level temperature and the actual r-values for single-glazed clear windows in case study #2.

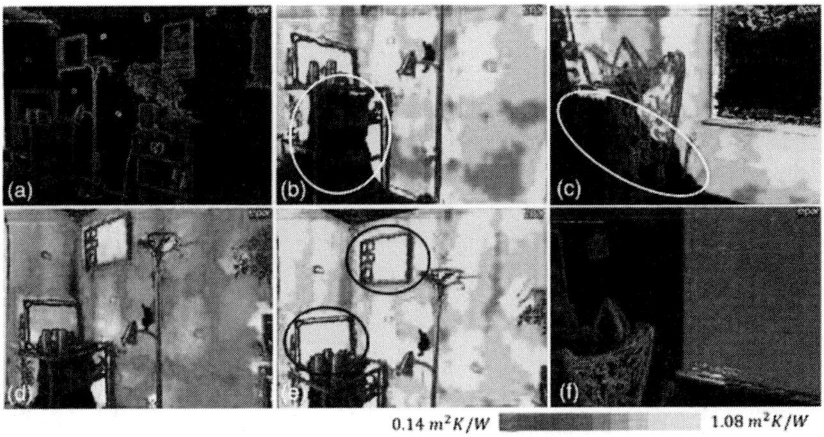

Figure 8. 3D VISUALIZATION ALONG WITH THE BUILDING GEOMETRY FOR CASE #1. The color spectrums shows variation between 0.14 to 1.08 m2K/W. **(A)**: 3D reconstructed built environments, **(B)** and **(C)**: thermal resistances of exterior walls with True Positive detection of occlusion, **(D)**: 3D thermal distribution with True Positive detection of occlusion, **(E)**: thermal resistances of exterior walls with False Negative detection of occlusion, and **(F)**: thermal resistances of a window component.

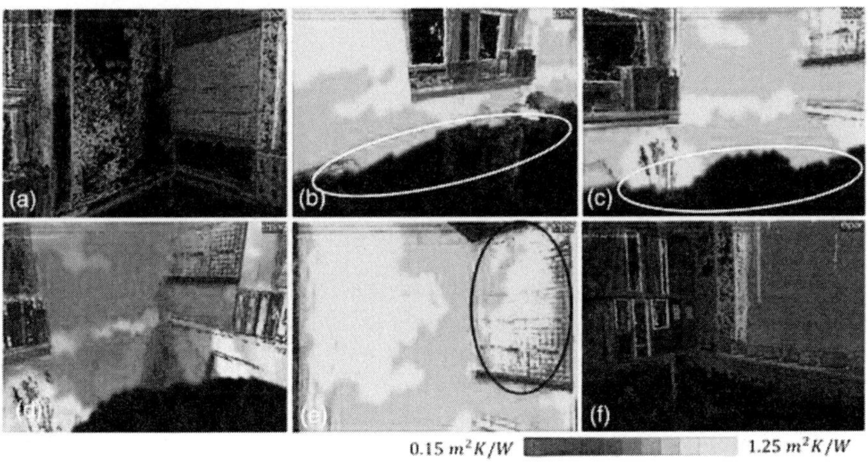

Figure 9. 3D VISUALIZATION ALONG WITH THE BUILDING GEOMETRY FOR CASE #2. The color spectrums shows variation between 0.15 to 1.25 m2K/W. **(A)**: 3D reconstructed built environments, **(B)** and **(C)**: thermal resistances of exterior walls with True Positive detection of occlusion, **(D)**: 3D thermal distribution with True Positive detection of occlusion, **(E)**: thermal resistances of exterior walls with False Negative detection of occlusion, and **(F)**: thermal resistances of a window component.

potential to improve the current practices by estimating the size of the building areas associated with condensation issues. In Figure 10, building surface areas that are likely to suffer from condensation issues are visualized in 3D ((a) in Case #1 $\cong 0.0046m^2$, (b) and (c) in Case #2 $\cong 0.0159m^2$). These results are consistent with inspection observations conducted at the time of data collection. When thermal images were captured, the surface temperatures of these areas were lower than the dew point. As shown in both cases, the condensation issues typically starts from around windowsills. To minimize the impact of these detected condensation problems, homeowners can (1) reduce the dew point temperature by decreasing the relative humidity or the indoor air temperature; (2) replace or add new insulation layers to increase the surface temperature on the areas where insulation layers have been missed or degraded; and finally (3) adjust the ventilation fan or better benefit from natural ventilation to increase the wind velocity around the problematic areas.

Figure 10 3D VISUALIZATION OF THE BUILDING AREAS IDENTIFIED WITH POSSIBLE CONDENSATION ISSUES (THE COLOR-CODING IS BASED ON THE METAPHOR OF TRAFFIC LIGHT COLORS) (A): CASE #1, (B) AND (C): CASE #2.

The timing of an inspection would impact the capability of detecting condensation issues. To continuously monitor changes in the environmental conditions and study their impacts on condensation problems, one can place environmental sensors on building interior surfaces and continuously monitor the relative humidity and also the surface and inside air temperatures. However, in practice, only a few environmental sensors could be placed per element. As such, considering the variations of thermal performance over surface areas, it is difficult to ensure if the surface condition measured by a few sensors is representative of the entirety of the building assemblies. Here, the proposed method has potential to provide a solution for selecting the proper locations of installing a limited number of environmental sensors for detecting condensation problems. As a complementary step, after detecting the condensation problems from the proposed one-time thermography-based method, the environmental sensors can be placed at the detected areas for continuous monitoring. This will allow proper positioning of a limited number of sensors and improves the efficiency of monitoring.

DOMAIN EXPERT EVALUATION AND DISCUSSIONS

To discuss the perceived benefits of the proposed method, face-to-face and phone interviews were conducted with eleven actual domain experts. The subjects had 2 to 10 years of practical experiences in energy auditing of existing buildings. They were first asked about the current best practices and bottlenecks in thermographic inspections for energy diagnostics in existing buildings. All subjects responded that today, thermography captured from existing buildings is considered the most effective tool for recording and communicating the as-is energy performance conditions to building owners. However, thermography is primarily used for qualitative documentations in support of leveraging the Home Energy Rating System (HERS) for reporting purposes at a later stage. Most responders used the word "qualitative" to describe the current process. For example, they noted that "Audits can take a long time, up to several hours for large residential houses. IR imaging is done piece-wise through a building. It takes detailed documentation to know where each image was taken and what it is supposed to be showing. This can be confusing, and is often detrimental to the client's understanding, even with a good service provider explaining discoveries.", "Sometimes it is hard to tell where the thermal image is coming from in the building. Often, our audits last about 2–4 hours. In this time, it is hard to note which image was taken where within the building."

Figure 11 illustrates the summary of the responses from the Likert questionnaires on the usefulness of each feature (1 = poor, 5 = excellent). As shown, participants generally felt that each features in the proposed work would be helpful for reducing the bottlenecks in current practices on thermography-based building condition assessment, e.g., "This is a very interesting step in the direction of recalling and identifying locations of leakage. This visualization tool will dramatically increase the potential for accurate assessments of issues.", "This visualization tool can be very efficient in detecting leaks in building insulation.", "In the summary, it would be useful to see an entire room in 3D to capture the proposed visualization tool's differentiating usefulness that cannot be done with 2D imaging."

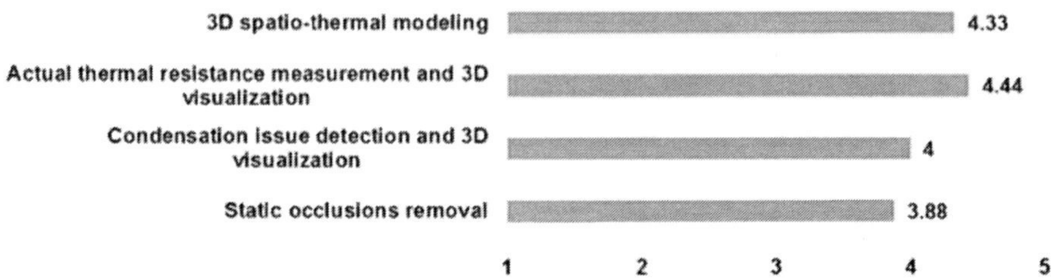

Figure 11 SUMMARIZED RESPONSES (AVERAGED OVER ALL SUBJECTS) FROM THE LIKERT SCALE QUESTIONNAIRES ABOUT THE USEFULNESS OF EACH FEATURE (1 = POOR, 5 = EXCELLENT).

The survey results show the potential of the proposed 3D visualization method for assessing the as-is building conditions as an intuitive communication tool. Since the main purpose of the proposed method is to provide guidance for making efficient and effective energy retrofit decisions by diagnosing and visualizing the as-is building conditions, the proposed method only needs to be conducted once per regular building diagnostic process (e.g., annual building inspection) and does not need to be performed more frequently. Hence, it does not add significant burden to the building practitioners, and allows the proposed method to be repeatable in parallel with thermographic inspection. The proposed method can inspect very small surfaces with potential performance problems (in fact, at the vertex-level of a mesh). This level of granularity, for example, would be helpful when installing blown-in loose-fill insulation is considered as one of the retrofit alternatives to improve the thermal resistance of the defected areas. However, if the detected areas are too small, retrofitting such problems is not likely to be practically feasible due to technical challenges and long payback period. Thus, practical retrofit decisions need to be taken up to a certain minimum size of retroffitable surface. The proposed method chooses not to dictate the minimum size of the retrofittable building area. Rather, by presenting the measurements at the point-level, the proposed method lets the building practitioners set their desired value prior to assessing different energy efficiency retrofit alternatives.

Despite such perceived and potential benefits, several participants noted that there is still room for improvement to alleviate bottlenecks in current practices, which are remaining open:

1. *Difficulty in in-situ measurement of as-is thermal properties for building assemblies*: Few participants had concerns about the difficulty in accurate measurement of thermal properties in building environments. Here, following prior works (Albatici and Tonelli [2010]; Dall'O' et al.

[2013]; Fokaides and Kalogirou [2011]; Madding [2008]), the deviation between the measurements and the notional values was examined. It is noted that wall assembles are not the best choice for such experiments since wall assemblies typically have multiple layers and are likely to suffer from non-uniform and time-variant degradations. Hence, in this case, using the notional R-values as the ground truth for the as-is thermal resistances may not be accurate. In contrary to walls, since windows are mostly standardized and their thermal properties are more likely to be consistent over time, their actual thermal resistance is less likely to deviate from the notional values derived from the as-built documents. Thus, window components were selected for experiments. It is not a trivial task to accurately estimate the thermal properties of the old building assemblies such as the case studies of this paper. Here, the estimation is based on (ColoradoENERGY [2013]; The Engineering Toolbox [2013]) which provide recommendations for such R-values and (European Standard (EN) [1997]). As a result for single-glazed clear windows, the averaged deviations were 12.81% (= (0.301-0.260)/0.301) for case #1 and 10.37% (= (0.301-0.274)/0.301) for case #2. Following the environmental assumptions presented in (Albatici and Tonelli [2010]; Dall'O' et al. [2013]; Fokaides and Kalogirou [2011]; Madding [2008]), the findings in the experiments are consistent with what was reported in those literatures (Section Thermal resistance (R-value) of building assemblies). Here, systematic errors on measuring the input variables for Equation (5) need to be considered. If the thermographic inspection is conducted in outdoor environments, the measurement accuracy would be affected by various external factors such as wind, sun radiation, and shadows. Since the temperatures are measured from inside surfaces and crumpled aluminum foils located in indoor environments, the accuracy of measuring those input variables is typically the same as it is claimed in the technical specification of the thermal camera. As can be seen in the Table 1, the measurement accuracy of the consumer-level thermal camera used is 2°C or ±2% of reading. Here, another prevalent challenge for accurate measurement in thermal resistances in a non-destructive manner is that the measurement needs to be conducted under steady-state conditions as discussed in Section (Environmental Assumptions and Experimental Setups). Despite significant efforts to form a steady-state heat transfer condition during thermographic inspection (e.g., the timing of data collection and experimental setups prior to image data collection as described in Section(Environmental Assumptions and Experimental Setups), in practice, it is not trivial to maintain a perfect steady-state condition during thermogaphic inspection. More research needs to be conducted on how a steady state condition of heat transfer can be continued during thermographic inspection and how the measurement errors caused by non-steady-state conditions can be accounted for and minimized for more accurate analysis;

2. *Continuous measurement to form time-series of thermal resistances*: Besides the measurement accuracy issues, a few responders described `4D visualization' as a desired feature, e.g. "Maybe a 4D visualization done with the presence of all the objects that exist in the building.", "A time series IR webcam of a building over the course of a year." In prior work on measuring actual U-values (Fokaides and Kalogirou [2011]), the results were validated based on different sources of information: instantaneous surface temperature data samples of 10 measurements obtained from a thermal camera and continuous time-series heat flux data collected from HFMs over a period of 168 hours. According to the experimental results, it is shown that the outcomes of the two methods were very close to each other in terms of their accuracies. However, if surface temperature data can be collected continuously, the actual heat resistances can be modeled and explored in 4D (3D + time) as opposed to static information. Visualizing dynamic energy performance metrics in form of a 3D animation can significantly improve the understanding of time-varying heat transfer phenomena. This also enables to provide an opportunity to analyze the impacts of building envelope retrofit decisions. For example, energy auditors can conduct a before/after analysis on thermal resistances when additional insulation layers are being installed. Such analyses would support implementation of periodic building maintenance programs. More research on 4D visualization of energy performance metrics need to be conducted; and finally

3. *Minimizing false negative detection for static occlusions*: One survey participant noted that "Accounting for furniture and other obstructions seems like a challenging task". The proposed method of setting a geometrical distance threshold has potential to ignore those building areas with static occlusions. Nevertheless, as discussed in the previous section, false negatives can still occur due to incomplete segmentations. In the experiments, those happen around thin objects on walls such as frames (e.g., Figures 8 and 9(e)) or tiny objects around windows (Figure 10(b) and (c)). This is because the distance between thermal points representing such objects in 3D point clouds and the nearest vertex in a geometrical baseline mesh is most likely to be within the threshold based on the registration error. As a result, the thermal resistance of the occluded building assemblies was not accurately calculated. More research needs to be conducted to better segment non-relevant objects from the reconstructed 3D scenes.

CONCLUSIONS

Effective building diagnostic tools should represent abnormal building conditions in an intuitive format to allow energy auditors better communicate the as-is building conditions to the homeowners. Toward this goal, this paper presents a method for visualizing the as-is building condition assessment at the level of 3D points. The outcome of the analyses are visualized within a 3D environment, and static occlusions are accounted for. In this 3D environment, practitioners can navigate and interactively query information about the building geometry, surface temperature data, thermal resistances, and analyze condensation problems. The proposed method was tested in a residential building and an instructional facility, and the following domain expert evaluation was conducted. By converting surface temperature data sensed from a consumer-level single thermal camera into 3D visualization of the actual R-value distributions and the location information of possible condensation problems, building practitioners can better understand the as-is building conditions. Future works include addressing the following challenges in (1) forming and maintaining a steady state condition of heat transfer; (2) studying 4D visualization of energy performance metrics based on time-series of building surface temperatures; (3) minimizing false negative segmentations for static occlusions; and (4) Varying the level of visualization for both global and local inspection of as-is energy performance based on different Levels of Detail (LOD). These are currently being explored as part of ongoing research.

DECLARATIONS

Competing interests

The authors declare that they have no competing interests.

Authors' contributions

YH designed and implemented the proposed 3D visualization of as-is building conditions, carried out the thermographic experiments and domain expert evaluations, and drafted the manuscript. MGF supervised the entire processes of this study, co-designed the 3D visualization tool, and edited the manuscript. All authors read and approved the final manuscript.

REFERENCES

1. Aelenei D, Henriques FMA: **ANALYSIS OF THE CONDENSATION RISK ON EXTERIOR SURFACE OF BUILDING ENVELOPES.** Energy and Buildings 2008, **40**(10):1866–1871.

2. Albatici R, Tonelli AM: **INFRARED THERMOVISION TECHNIQUE FOR THE ASSESSMENT OF THERMAL TRANSMITTANCE VALUE OF OPAQUE BUILDING ELEMENTS ON SITE.** Energy and Buildings 2010, **42**(11):2177–2183.

3. Bellia L, Minichiello F: **A SIMPLE EVALUATOR OF BUILDING ENVELOPE MOISTURE CONDENSATION ACCORDING TO AN EUROPEAN STANDARD.** Building and Environment 2003, **38**(3):457–468.

4. Borrmann D, Elseberg J, Nüchter A: **THERMAL 3D MAPPING OF BUILDING FAÇADES.** In Intelligent Autonomous Systems 12. Edited by: Lee S, Cho H, Yoon K-J, Lee J. Springer Berlin Heidelberg, Berlin, Germany; 2013:173–182.

5. ColoradoENERGY. (2013). R-Value Table: Insulation Values For Selected Materials. http://www.coloradoenergy.org/procorner/stuff/r-values.htm.

6. Dall'O' G, Sarto L, Panza A: **INFRARED SCREENING OF RESIDENTIAL BUILDINGS FOR ENERGY AUDIT PURPOSES: RESULTS OF A FIELD TEST.** Energies 2013, **6**(8):3859–3878.

7. European Standard (EN). (1997). Glass in Building - Determination of Thermal Transmittance (U-value) - Calculation Method (EN 673).

8. FLIR system. (2010). Thermographhic Measurement Techniques (FLIR Reporter Professional, Professional Edition, pp. 133–137).

9. Fokaides PA, Kalogirou SA: **APPLICATION OF INFRARED THERMOGRAPHY FOR THE DETERMINATION OF THE OVERALL HEAT TRANSFER COEFFICIENT (U-VALUE) IN BUILDING ENVELOPES.** Applied Energy 2011, **88**(12):4358–4365.

10. Golparvar-Fard M, Ham Y: **AUTOMATED DIAGNOSTICS AND VISUALIZATION OF POTENTIAL ENERGY PERFORMANCE PROBLEMS IN EXISTING BUILDINGS USING EPAR -ENERGY PERFORMANCE AUGMENTED REALITY-MODELS.** Journal of Computing in Civil Engineering 2014, **28**(1):17–29.

11. Ham Y, Golparvar-Fard M: **AN AUTOMATED VISION-BASED METHOD FOR RAPID 3D ENERGY PERFORMANCE MODELING OF EXISTING BUILDINGS USING THERMAL AND DIGITAL IMAGERY.** Advanced Engineering Informatics 2013, **27**(3):395–409.

12. Ham Y, Golparvar-Fard M: **EPAR: ENERGY PERFORMANCE AUGMENTED REALITY MODELS FOR IDENTIFICATION OF BUILDING ENERGY PERFORMANCE DEVIATIONS BETWEEN ACTUAL MEASUREMENTS AND SIMULATION RESULTS.** Energy and Buildings 2013, **63:** 15–28.

13. Ham, Y, & Golparvar-Fard, M. (2014). Three-dimensional thermography-based method for cost-benefit analysis of energy efficiency building envelope retrofits. Journal of Computing in Civil Engineering., [http://ascelibrary.org/doi/abs/10.1061/%28ASCE % 29CP.1943-5487.0000406]

14. Hoff R: Water condensation – a Temperature Related Phenomenon. 2011. http://www.moistureview.com/blog-034.

15. Horn B: **CLOSED-FORM SOLUTION OF ABSOLUTE ORIENTATION USING UNIT QUATERNIONS.** Journal of the Optical Society 1987, **4**(4):629–642.

16. International Organization for Standardization (ISO). (2007). Building Components and Building Elements-Thermal Resistance and Thermal Transmittance-Calculation Method (ISO 6946).

17. Lagüela S, Díaz-Vilariño L, Martínez J, Armesto J: **AUTOMATIC THERMOGRAPHIC AND RGB TEXTURE OF AS-BUILT BIM FOR ENERGY REHABILITATION PURPOSES.** Automation in Construction 2013, **31:** 230–240.

18. Madding R: Finding R-Values of Stud Frame Constructed Houses with IR Thermography. Proc, InfraMation; 2008.

19. Mikron Instrument Company. (2012). Table of Emissivity of Various Surface.

20. Roth K, Westphalen D, Feng M, Llana P, Quartararo L: The Energy Impact of Commercial Building Controls and Performance Diagnostics: Market Characterization, Energy Impact of Building Faults and Energy Savings Potential. TIAX LCC Technical Report Cambridge, MA; 2005.

21. Singham JR: **TABLES OF EMISSIVITY OF SURFACES.** International Journal of Heat and Mass Transfer 1962, **5**(1–2):67–76.

22. The Engineering Toolbox. (2013). Heat Loss through Building Elements due to Transmission. http://www.engineeringtoolbox.com/heat-loss-transmissiond_748. html.

23. U.S. DOE. (2010). 2010 U.S. DOE Buildings Energy Databook. Washington D.C.: U.S. Department of Energy.

24. Wang C, Cho Y, Gai M: **AS-IS 3D THERMAL MODELING FOR EXISTING BUILDING ENVELOPES USING A HYBRID LIDAR SYSTEM.** Journal of Computing in Civil Engineering 2013, **27**(6):645–656.

25. Wanielista M, Kersten R, Ron E: Hydrology Water Quantity and Quality Control. John Wiley & Sons, Hoboken, NJ; 1997.

11

CHAPTER

PERFORMANCE EVALUATION OF HEAT EXCHANGERS IN OTEC USING AMMONIA/WATER MIXTURE AS WORKING FLUID

Takafumi Morisaki, Yasuyuki Ikegami

Institute of Ocean Energy, Saga University, Saga, Japan

ABSTRACT

The ocean thermal energy conversion (OTEC) system is a promising solution to provide stable electricity supply. Although the available temperature difference in OTEC systems is small, an ammonia/water mixture as working fluid is expected to decrease irreversible losses in the heat exchangers and to improve system performance. However, in actual heat exchangers, an adequate temperature crossing does not occur in the condenser but in the evaporator. Therefore, clarification of this characteristic is important. To date, the logarithmic temperature difference (LMTD) method is used in performance evaluations of OTEC heat exchangers. This method is of limited use if physical properties of fluids vary. A generalized mean temperature difference (GMTD) method is introduced to perform this evaluation. As changes in fluid property values can be considered in the GMTD method, method dependencies on heat exchanger characteristics, effectiveness, and system characteristics can be studied. In particular, GMTD and LMTD using a pure substance were found to be almost equal. Mean temperature differences using mixtures as working fluid were higher in the evaporator, but lower in the condenser, from the GMTD method than from the LMTD method. For higher ammonia concentrations in ammonia/water mixtures, the mean temperature differences from both methods are different.

Keywords: Ammonia/Water Mixture; Evaporation and Condensation Pressure; OTEC; GMTD; LMTD

1. INTRODUCTION

Energy and environmental problems are becoming more critical; the development of natural energy is desired. Ocean thermal energy conversion (OTEC) using temperature differences between warm surface seawater and cold deep seawater can supply stable electrical power [1] integrated into a variety of applications, such as seawater desalination, residential cooling, and aquaculture [2-6]. In addition, the beneficial effects of OTEC systems on global warming and energy shortage problems seem promising. Oceans have a huge amount of stored thermal energy, although, the energy density is low. As an OTEC system operates in conditions of small temperature differences, an appropriate working fluid must have a low boiling point if it is to be used in a closed-cycle OTEC. The OTEC system requires warm water, cold seawater, and a working fluid pump power; the ratio of the warm and cold water pump power to output power is comparatively higher than that of conventional fossil power plants.

Research is continuing on the Rankine cycle which is the conventional closed-cycle process exploiting lowboiling-point working fluids such as ammonia. Using a pure substance as working fluid, the thermal efficiency of the OTEC cycle increases with decreasing irreversible losses in the cycle or increasing effective temperature difference, specifically the difference between evaporating and condensing temperatures of the working fluid. For that reason, it is necessary to decrease energy losses in the heat exchangers, namely to decrease the temperature difference between seawater and working fluid. This is done by increasing the heat transfer area or the overall heat transfer coefficient. Using non-azeotropic mixtures such as ammonia/water as working fluids in the OTEC cycle was proposed by Kalina in 1982 [7]. Uehara et al. [8] indicated that reducing the irreversible losses in heat exchangers using non-azeotropic mixtures leads to improvements in the thermal efficiency of the cycle. Furthermore, they focused on the Kalina cycle, and presented a new cycle using the ammonia/water mixture [9]. Several theoretical and experimental investigations have reported that thermal efficiency of such Kalina cycles is higher than that of the Rankine cycle using pure ammonia [10-15]. Other studies have reported that the effective temperature difference and heat transfer coefficient decrease with changes in concentration in the boundary layer during evaporation and condensation [16-18]. Previous studies have investigated the operating conditions and equipment of an experimental apparatus using ammonia/water mixtures to improve system performance. Ikegami et al. [19] performed experiments to investigate the influence of system performance on ammonia concentration in the mixture. The experiments were performed to assess the stability of the ammonia/water mixture power cycle [20]. The results from these experiments showed that it is critically important to enhance the performance of the condenser in increasing the performance of the total OTEC system. Furthermore, the characteristics of the ammonia/water mixture cycle have not as yet been thoroughly clarified experimentally.

To enhance the ammonia/water mixture cycle, clarification of the heat exchanger performance is extremely important. The traditional and effectual evaluation method of heat exchanger performance is the logarithmic mean temperature difference (LMTD) method but application of this method is limited to heat exchangers using working fluids having constant heat transfer coefficient and thermal properties. The generalized mean temperature difference (GMTD) method however enables variations in thermal properties in heat exchange processes to be included. Utamura et al. [21] first proposed the GMTD method, and applied it to evaluate the heat exchange process between hot water and supercritical carbon dioxide as a working fluid. As a result, validity of the GMTD method was proved. The GMTD is believed applicable in the evaluation of heat exchangers using ammonia/water although properties of the mixture are quite distinct. It is not clear whether the GMTD method is valid for heat exchangers using non-azeotropic mixtures as working fluid.

In this paper, clarification is obtained on the characteristics of heat exchanger using ammonia/water mixtures in improving cycle performance, achieved with the available methods of evaluation heat exchangers. A comparison is made between LMTD and the recently developed GMTD calculation methods as performed on the heat exchanger using ammonia/water mixture and the applicability of the GMTD method is assessed.

2. TEMPERATURE CHANGE IN HEAT EXCHANGERS WITH AMMONIA/WATER MIXTURE

In Figures 1(a) and (b), the distributions of the fluid temperature within the heat exchanger is given as a function of distance (Δx) and heat flow rate (ΔQ), respectively. The temperature variation of the ammonia/water mixture changes in the heat exchanger owing to the large difference between its boiling point and its dew point. Consequently, an increase in available exergy of the system is expected with this mixture as working fluid. From Figure 1(a), the improvement in heat transfer performance and the ratios of the working fluid temperature to the heat source temperature in the evaporator and to the cold In Figures 1(a) and (b), the distributions of the fluid

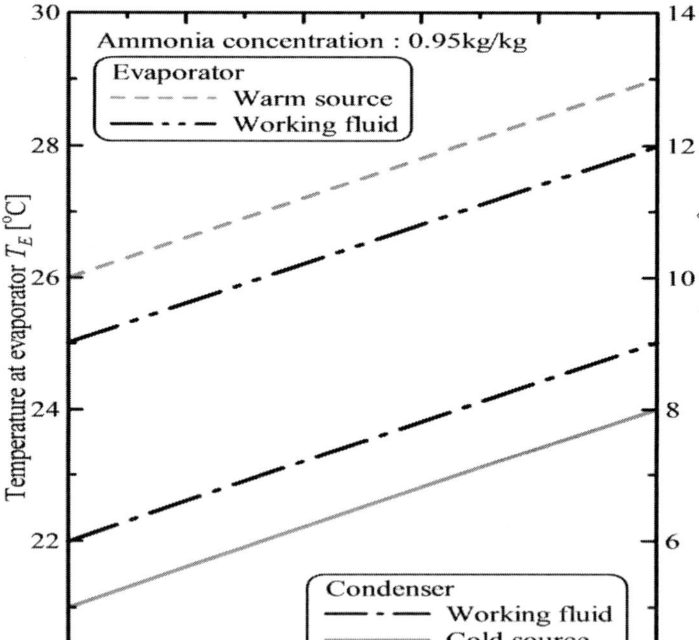

(a)

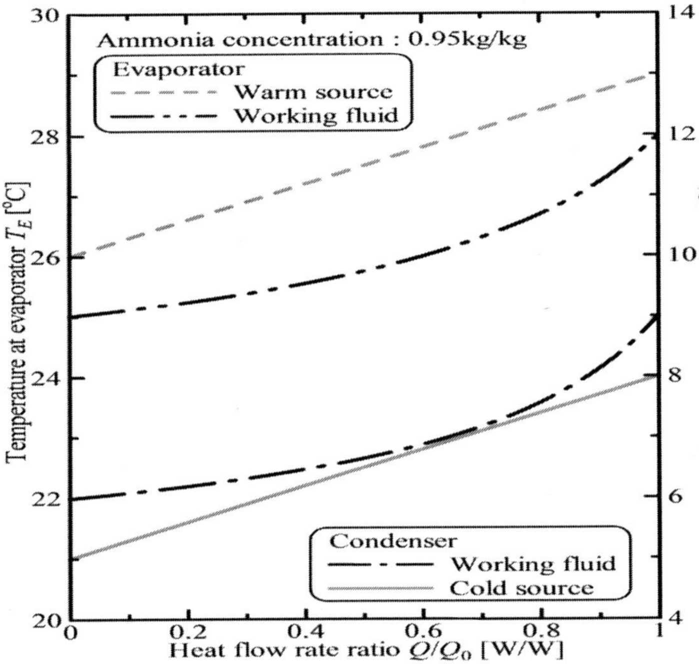

(b)

Figure 1. Temperature change in heat exchangers. Relationships between temperature and (a) length ratio and (b) heatflow-rate ratio for evaporator and condenser.

temperature within the heat exchanger is given as a function of distance (Δx) and heat flow rate (ΔQ), respectively. The temperature variation of the ammonia/water mixture changes in the heat exchanger owing to the large difference between its boiling point and its dew point. Consequently, an increase in available exergy of the system is expected with this mixture as working fluid. From Figure 1(a), the improvement in heat transfer performance and the ratios of the working fluid temperature to the heat source temperature in the evaporator and to the cold source temperature in the condenser are important when considering enhancing cycle performance. From Figure 1(b), the working fluid temperature is a convex curve of the heat flow rate. The working fluid temperature approaches the warm water temperature in the inlet and outlet of evaporator but approaches the cold water temperature in the middle of the condenser. The pinchpoint in particular is not well determined by the inlet and outlet temperatures in condenser. In addition, the traditional evaluation method underestimates heat transfer performance. Hence, the GMTD method, which accounts for the variation in the thermal properties, has been introduced for this performance evaluation. In contrast, for a pure ammonia working fluid, temperatures are constant in these heat exchangers because the thermal properties are stable.

2.1. Traditional Evaluation Method

The LMTD method has traditionally been used to evaluate the temperature difference between the heat source and the working fluid in a heat exchanger. The LMTD method is used only when the thermophysical properties of the local fluid in the heat exchanger and the overall heat transfer coefficient are constant. Hence, the fluid in the heat exchanger undergoes temperature changes as shown in Figure 2 with the horizontal axis conceptually representing the length. The LMTD method is expressed as follows:

$$\Delta T_{LMTD} = \frac{(T_{HI} - T_{LO}) - (T_{HO} - T_{LI})}{\ln\left[(T_{HI} - T_{LO})/(T_{HO} - T_{LI})\right]} \quad . (1)$$

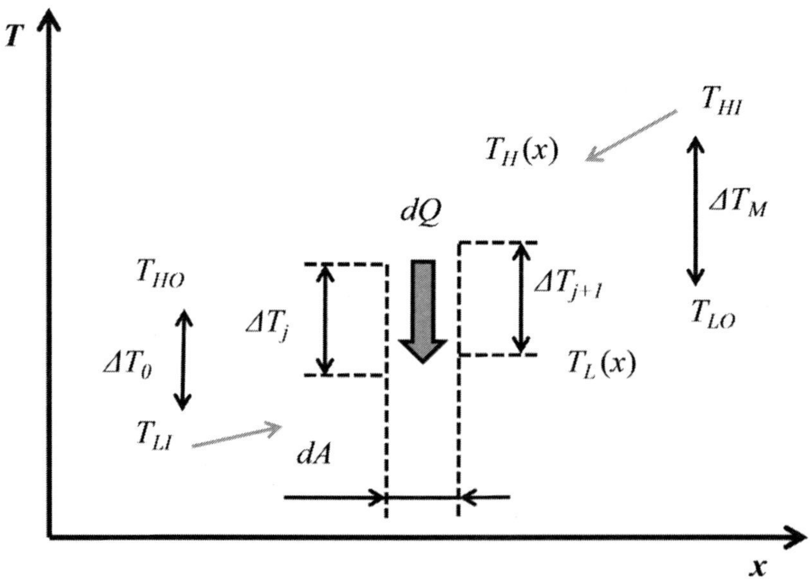

Figure 2. Fluid temperatures in the counter-flow heat.

where T_{HI}, T_{HO}, T_{LI} and T_{LO} are the high and the low temperature fluids heat exchanger inlet and outlet temperatures.

2.2. Innovated Evaluation Method

For an improved evaluation, it is necessary to consider variations in the thermophysical properties that better reflect conditions within actual heat exchangers. Therefore, the calculation results of LMTD are compared with those of GMTD and assessed. Using Figure 2, the heat flow rate for the heat exchanger, dQ from x to x + dx, is calculated using:

$$dQ = U_i dA_i \Delta T \quad . (2)$$

where U_i represents the overall heat transfer coefficient for the ith segment, mass flow rate of the warm or cold source, dA_i is the heat transfer surface area, and ΔT is the temperature difference between highand low-temperature fluids. Integrating equation over all segments of the exchanger yields

$$\int_0^{Q_0} \left[dQ / \Delta T(Q) \right] = \int_0^{A_i} U_i dA_i \quad . (3)$$

Then, the GMTD ΔT_{GMTD} is defined as follows [21]:

$$1/\Delta T_{GMTD} = \left(1/Q_0\right) \int_0^{Q_0} \left[dQ / \Delta T(Q) \right] \quad . (4)$$

The temperature integral is approximated as:

$$\int_0^{Q_0} \left[dQ / \Delta T(Q) \right] = \sum_{j=1}^{M} \left[2q / \left(\Delta T_j - \Delta T_{j+1} \right) \right] \quad . (5)$$

Therefore, the GMTD is calculated by:

$$\Delta T_{GMTD} = M \Bigg/ \sum_{j=1}^{M} \left[2 / \left(\Delta T_j - \Delta T_{j+1} \right) \right] \quad . (6)$$

3. Comparison between Pure Ammonia and Ammonia/Water Mixture

Figure 3 shows the correlation between GMTD and LMTD for the evaporator and the condenser calculated for both pure ammonia and ammonia/water mixture as working fluid. The evaporator and the condenser are counter-flow-type heat exchangers; both warm and cold sources are water; for the warm water, the inlet temperature is 28°C, and the outlet temperature is 25°C; for the cold water, the inlet temperature is 5°C, and the outlet temperature is 8°C. For pure ammonia, the evaporation temperature varied from 27.5°C to 26.0°C; the condensation temperature varied from 5.5°C to 7.0°C. For the ammonia/water mixture, the temperature at the evaporator inlet varied from 27.5°C to 26.0°C, and at the evaporator the outlet temperature varied from 24.5 to 23.0; the temperature at the condenser inlet varied from 6.0°C to 7.0°C, and at the condenser outlet varied from 8.5°C to 10.0°C; the ammonia concentration is 0.95 kg/kg. The working fluids are assumed to be saturated liquid at the evaporator inlet and outlet, and saturated vapor at the condenser outlet and inlet. The physical properties of ammonia are evaluated using the P-PROPATH database, and those of ammonia/water mixture are evaluated using the M-PROPATH database (Ibrahim and Kelein) [22].

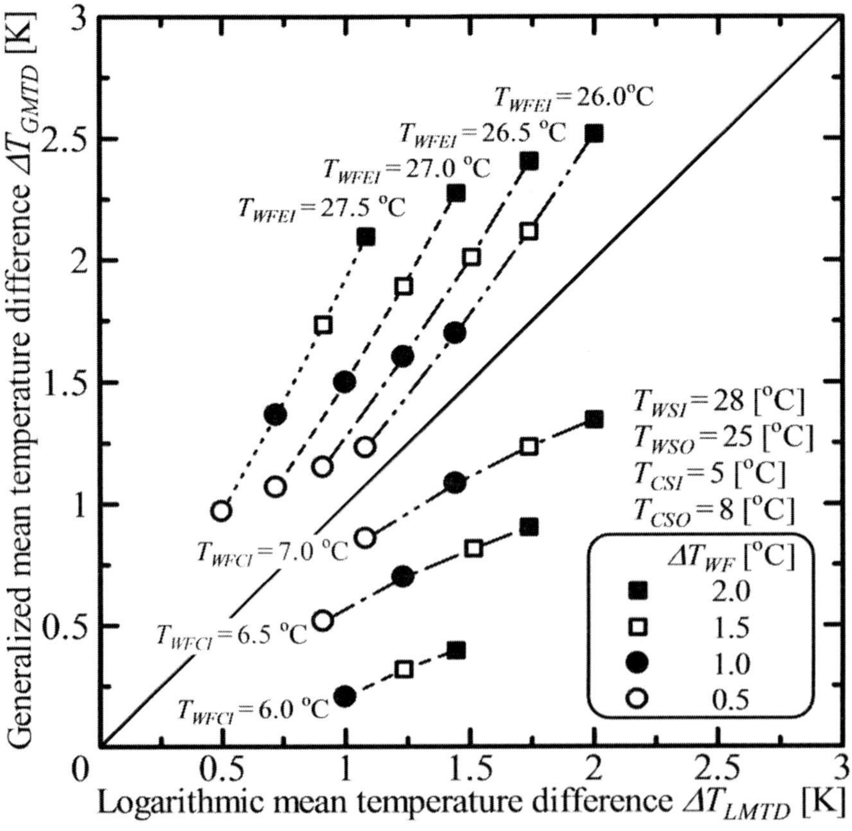

Figure 3. Temperature change in heat exchangers.

The mean temperature differences between pure ammonia and heat source from the LMTD method and GMTD method are almost equal, as seen in Figure 3. In contrast, the mean temperature difference between ammonia/water mixture and heat source is higher in the evaporator from the GMTD method than from the LMTD method, but lower in the condenser. The variation of working fluid temperature with heat-flow-rate ratio produces a convex downward curve, as seen in Figure 1(b). For the evaporator, the working fluid temperature approaches the warm water temperature at the evaporator inlet and outlet. The mean temperature difference is higher in the evaporator from the GMTD method than from the LMTD method. For the condenser, the working fluid temperature approaches the cold water temperature of the condenser. The mean temperature difference with the GMTD is lower than that with the LMTD method.

4. MASS-FLOW-RATE RATIO DEPENDENCIES FROM GMTD AND LMTD MODELS

Figure 4(a) shows the mean temperature difference between ammonia/water mixture and warm water from the LMTD and GMTD methods and the evaporation pressure P_E of the working fluid, in the evaporator with the massflow-rate ratio of working fluid to warm source. Here P_E is defined as the maximum evaporation pressure of the working fluid, P_{max} for these calculation settings. Figure 4(b) shows the mean temperature difference between the ammonia/water mixture and cold water for both LMTD and GMTD methods as well as the conden sation pressure of the working fluid, P_C, in the condenser against the mass-flow-rate ratio of working fluid to cold source. The condensation pressure, P_C is defined as the minimum condensation pressure of the working fluid, P_{min} for these calculation settings. The same settings for the inlet and outlet temperatures for evaporator and condenser described in Section 3 apply here. The ammonia concentration in the mixture is 0.95 kg/kg; the mass-flow-rate ratio of working fluid to heat source, γ varied from 1.0×10^{-3} to 5.0×10^{-2}. The temperature difference at the pinchpoint is 0.5°C. The working fluid is also assumed to be saturated liquid at the evaporator inlet and saturated vapor at the condenser outlet. The physical properties of ammonia/water mixture are evaluated using the M-PROPATH database [22].

The mean temperature difference is higher in the evaporator from the GMTD method than from the LMTD method at low flow-rate ratios (Figure 4(a)). The pinchpoint appears at the evaporator inlet and outlet, reflecting the convex downward nature of the variation of the working fluid temperature compared with the warm water temperature. The convexity of the working fluid temperature changes strongly when the temperature difference of the working fluid between the evaporator inlet and outlet increases. The maximum pressure decreases with decreasing mass-flow-rate ratio at low flow-rate ratio, that is, 1.64×10^{-2} or less under these calculation conditions, because of the increasing working fluid temperature difference between evaporator inlet and outlet. In contrast, the mean temperature differences from GMTD and LMTD methods are almost equal at the high flow-rate ratios. The convexity of the working fluid temperature is not strong at the high flow rate ratio, that is, 5.0×10^{-2} or more under these calculation settings, because the working fluid temperature difference between evaporator inlet and outlet is small.

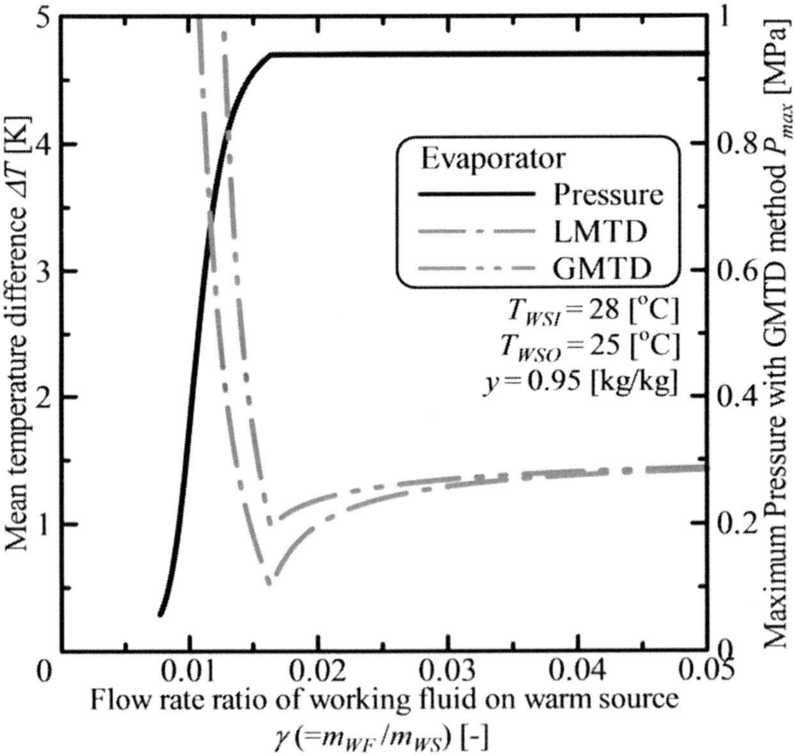

(a)

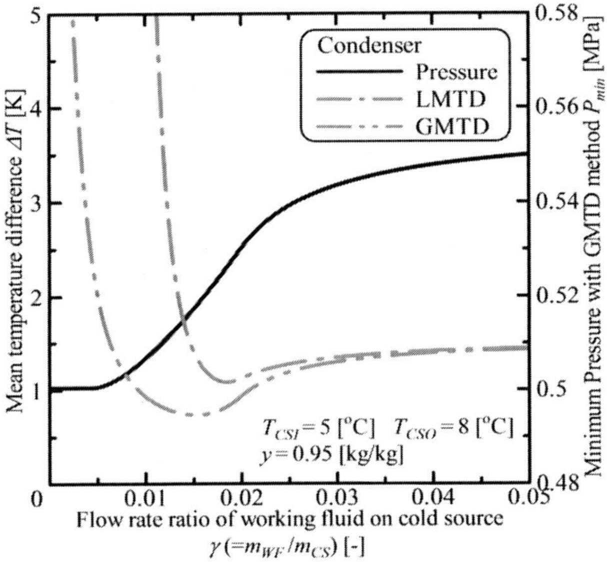

(b)

Figure 4. Temperature change in heat exchangers. (a) Evaporator; (b) Condenser.

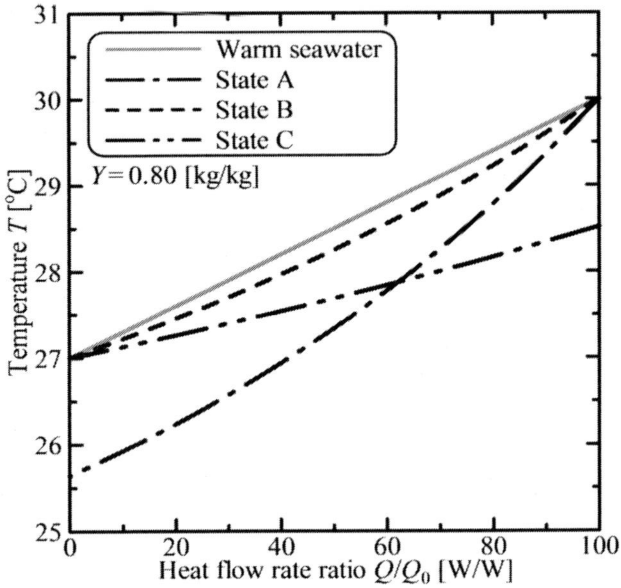

(a)

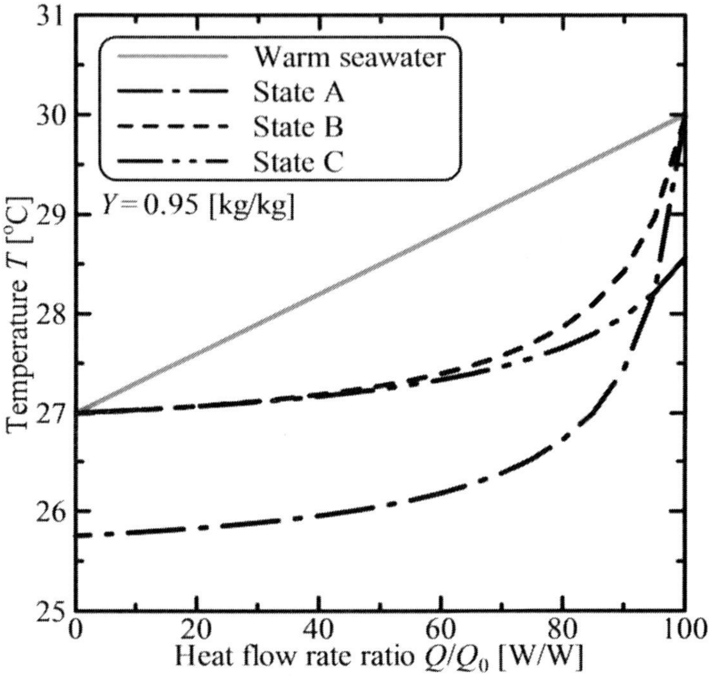

(b)

Figure 5. Temperature change in the evaporator. (a) Y = 0.80 kg/kg; (b) Y = 0.95 kg/kg.

Figures 5(a) and (b) shows an example where there are variations in the working fluid and warm water temperatures in the evaporator at heat-flow-rate ratio of 0.80 kg/kg and 0.95 kg/kg ammonia concentration, respectively, and the temperature difference at the pinch-point is 0 K. The working fluid temperature curve against heat flow rate is convex downward. Its slope rises gradually as the ammonia concentration in the mixture increases. The path between the working fluid outlet temperature and warm source inlet temperature is defined as "State A". The mass-flow-rate ratio for "State A" is under 1.64×10^{-2}. A similar path at the evaporator inlet and outlet, defined as "State B", has mass-flow-rate ratio of 1.64×10^{-2}, and the mean temperature differences from the GMTD and LMTD methods are minimal. The path between the working fluid inlet temperature and warm source outlet temperature is defined as "State C". The variation in the maximum pressure is constant because at evaporator inlet the working fluid is saturated liquid. The mass-flow-rate ratio for "State C" exceeds 1.64 × 10^{-2}.

The mean temperature difference is lower from the GMTD method than from the LMTD method at low flow-rate ratios in the condenser (Figure 4(b)). For the condenser, the working fluid is high temperature fluid, and the cold water is low temperature fluid. A pinchpoint appears as a consequence of convex downward variation in the working fluid temperature compared with that for the cold water temperature. Similar to the convexity in the working fluid temperature in the evaporator, the temperature in the condenser changes when the temperature difference of the working fluid between the condenser inlet and outlet increases. Therefore, the minimum pressure decreases gradually with decreasing mass-flowrate ratio. The mean temperature differences from the GMTD and LMTD methods are almost equal at highflow-rate ratios, that is, 5.0 × 10^{-2} or more with these calculation settings, because the working fluid temperature difference between evaporator inlet and outlet is small.

Figures 6(a) and (b) show as a further example with property variations in the working fluid, the cold water temperatures in the condenser against heat-flow-rate ratio at ammonia concentration of 0.80 kg/kg and 0.95 kg/kg, respectively, when the temperature difference of the pinch-point is 0 K. The path between the working fluid outlet temperature and cold source inlet temperature is defined as "State A". The variation of the minimum pressure is constant from the assumption that the working fluid is saturated liquid at the condenser outlet. The mass-flow-rate ratio for "State A" is 4.4 × 10^{-3} or less. Similarly, the path between the working fluid outlet temperature and warm source temperature in the condenser is defined as "State B". The pinch-point for "State B" is transferred from the cold water condenser outlet to the inlet when the mass-flow-rate ratio decreases. The mass-flow-rate ratio for "State B" is from 4.5 × 10^{-3} to 1.96 × 10^{-2}. The mean temperature difference from the GMTD method is a minimum when the mass-flow-rate ratio is 1.53 × 10^{-2}, but the minimum from the LMTD method occurs when the mass-flow-rate ratio is 1.85 × 10^{-2}. The path between the working fluid inlet temperature and warm source outlet temperature is defined as "State C". From the GMTD method, the mean temperature difference is lower for "State B" in the condenser. The mass-flow-rate ratio for "State C" is 1.97 × 10^{-2} or more.

5. AMMONIA CONCENTRATION AND COLD SOURCE TEMPERATURE DEPENDENCIES FROM GMTD AND LMTD MODELS

Obtained using the GMTD method, Figures 7(a) and (b) show the dependence of the maximum pressure in the evaporator on the mass-flow-rate ratio of working fluid to warm source. Similarly, Figures 8(a) and (b) show the dependence of minimum pressures for the condenser on the mass-flow-rate ratio of working fluid to cold source. The same calculation settings as describe above in Section 3 are used. The mass-flow-rate ratio of the working fluid with heat source varied from 1.0×10^{-3} to 5.0×10^{-2}. Again, the working fluid is saturated liquid at the evaporator inlet and saturated vapor at the condenser outlet. The physical properties of ammonia/water mixture are evaluated using the M-PROPATH database [8]. The ammonia concentration in the mixture varied from 0.99 to 0.80 kg/kg in Figures 7(a) and 8(a). The warm water temperature at the evaporator outlet varied from 27°C to 24°C in Figure 7" target="_self"> Figure 7(b). The cold water temperature at the condenser outlet varied from 6°C to 9°C in Figure 8(b).

The convexity of the working fluid temperature changes strongly when the ammonia concentration in the mixture increases (Figures 5(a) and (b)). For "State A", the maximum pressure decreases gradually at low ammonia concentrations compared with high ammonia concentrations as mass-flow-rate ratio decreases. In contrast, the mass-flow-rate ratio for "State B" increases with increasing warm water temperature difference between evaporator inlet and outlet.

Like convexity for the working fluid temperature in the evaporator, that in the condenser changes when the ammonia concentration in the mixture increases, (Figures 6(a) and (b)). Therefore, the mass-flow-rate ratio for "State B" increases with decreasing ammonia concentration. The slope of the cold water temperature curve is steep when the cold water temperature difference between the condenser inlet and outlet increases. The range for "State B" increases as the cold water temperature difference increases because the pinch-point at "State B" is transferred from the cold water condenser outlet to the inlet.

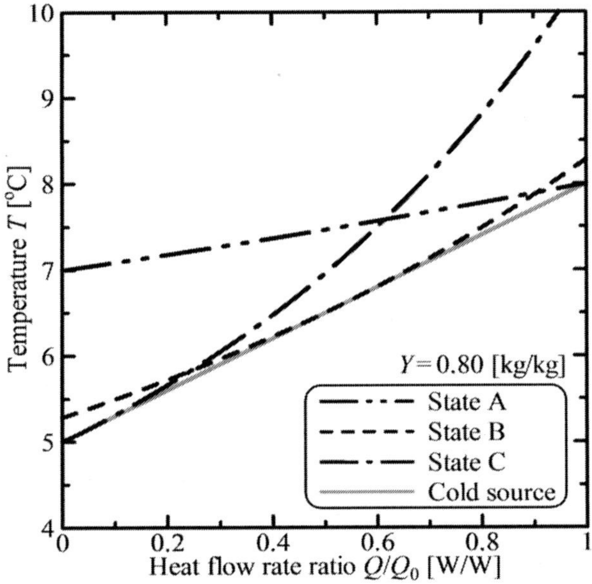

(a)

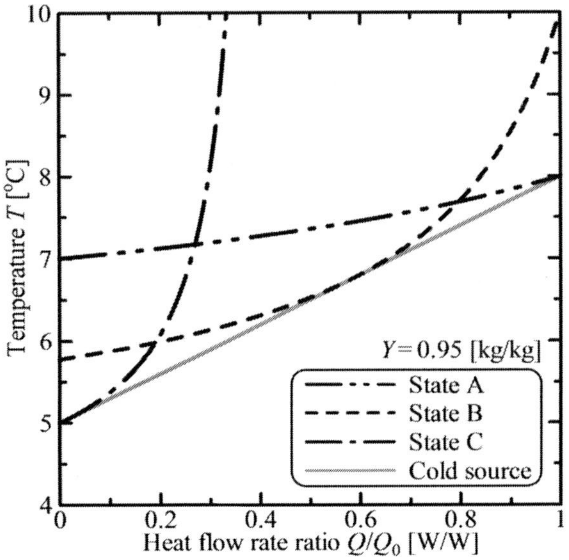

(b)

Figure 6. Temperature change in the condenser. (a) Y = 0.80 kg/kg; (b) Y = 0.95 kg/kg.

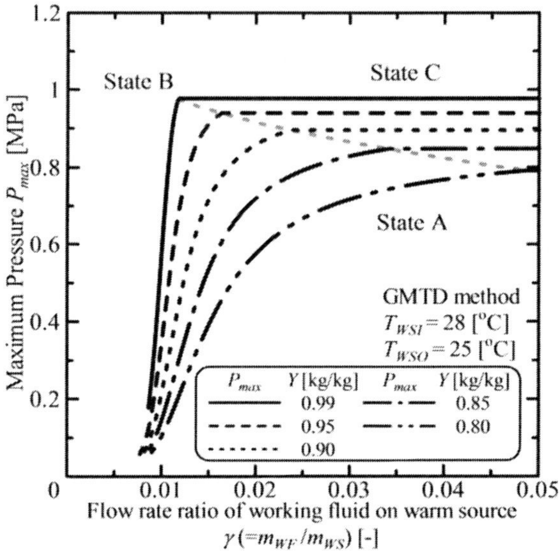

(a)

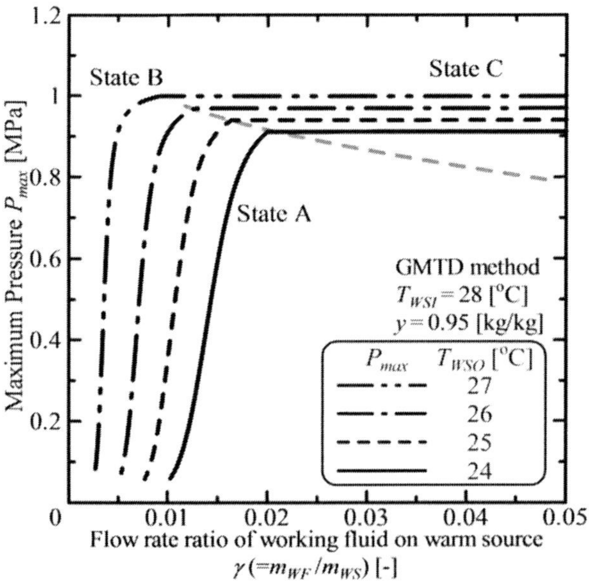

(b)

Figure 7. Maximum evaporation pressure. (a) Ammonia concentration; (b) Warm source outlet temperature.

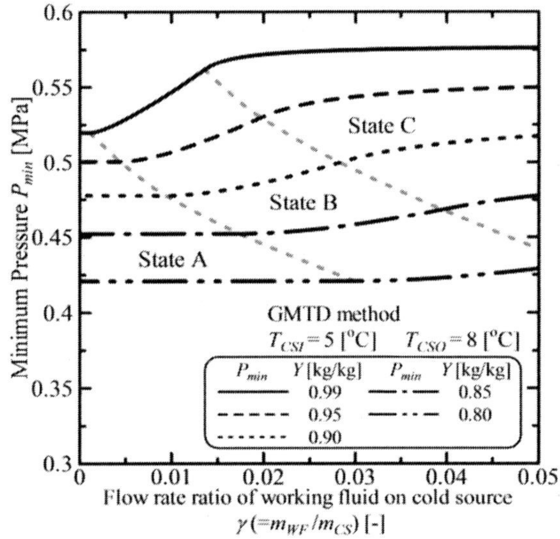

(a)

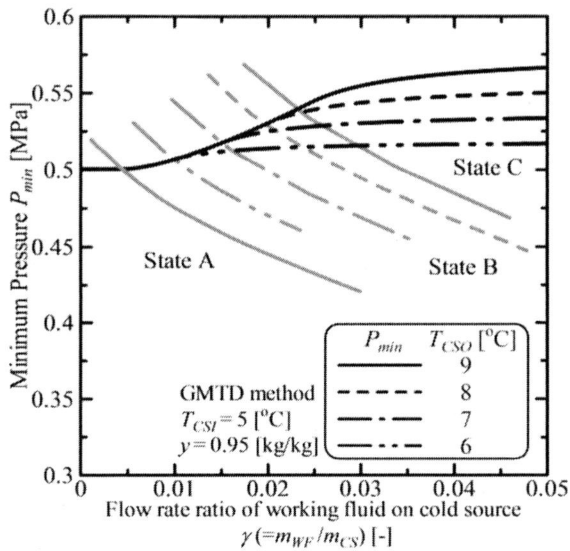

(b)

Figure 8. Maximum condensation pressure. (a) Ammonia concentration; (b) Cold source outlet temperature.

The correlation between the mean temperature differences from both GMTD and LMTD methods plotted in Figures 7(a) and 8(a) is presented in Figure 9; the broken line refers to an ammonia concentration of 0.99 kg/kg, the chain line to 0.90 kg/kg, and the two-dot chain line to 0.80 kg/kg; the circles refers to a mass-flow-rate ratio of 5.0 × 10^{-2}, the black circle 3.0 × 10^{-2}, and the rectangle 2.0 × 10^{-2}. The convexity of the working fluid temperature changes strongly when ammonia concentration in the mixture increases (Figures 5 and 6). Therefore, the mean temperature difference differs between the GMTD and LMTD methods when the ammonia concentration in the mixture increases.

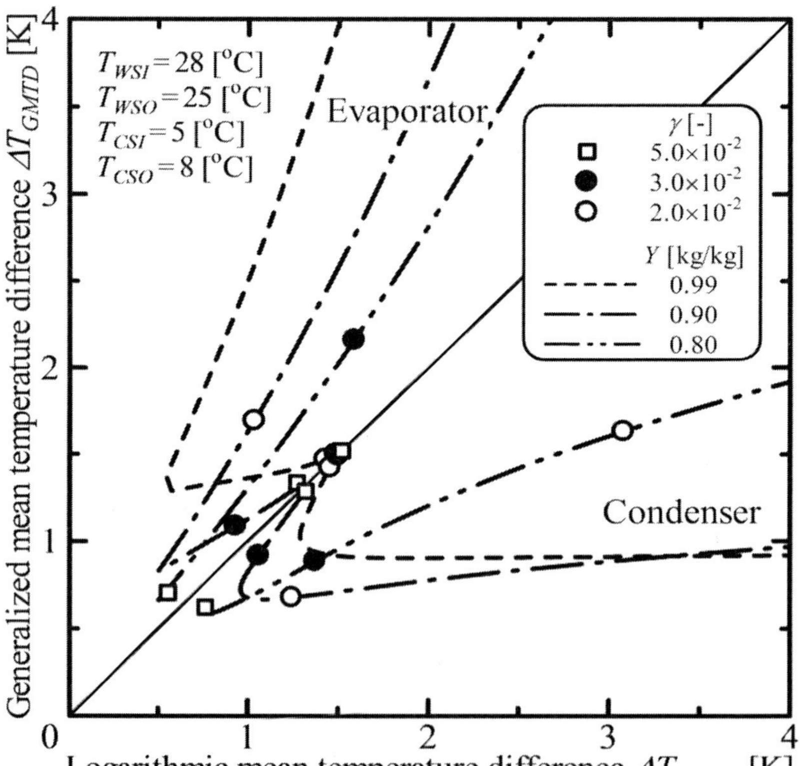

Figure 9. Relationship between GMTD and LMTD.

6. CONCLUSIONS

To obtain reliable data on the characteristics of a heat exchanger that uses ammonia/water mixture as working fluid, the recently developed GMTD method, which is to replace the traditional LMTD method, was assessed for validity in application. The conclusions are as follows:

1) Mean temperature differences between pure ammonia and water obtained with the LMTD and GMTD methods are almost equal. In contrast, the evaluation of heat exchange using ammonia/water mixtures using the LMTD method was unsatisfactory. Temperatures in the ammonia/water mixture differ from warm water temperatures in the evaporator, and only approach cold water temperatures in the condenser.

2) Mean temperature differences are higher in the evaporator and lower in the condenser from the GMTD method than from the LMTD method, because the convexity of ammonia/water mixture temperature changes strongly as ammonia concentration increases. From the GMTD method, the mean temperature difference is lower when the working fluid temperature approaches the cold water temperature in the condenser.

3) For "State A", the maximum pressure in the evaporator decreases with decreasing mass-flow-rate ratio of ammonia/water mixture to water, owing to an increase in the temperature difference between the evaporator inlet and outlet. In contrast, the variation in the maximum pressure is constant at "State C" because the working fluid is assumed to be saturated liquid at evaporator inlet. The maximum pressure decreases with decreasing ammonia concentration in the mixture.

4) The variation in the minimum pressure in the condenser is constant assuming that the working fluid at condenser outlet is saturated liquid. The pinch-point for "State B" is transferred from the cold water condenser outlet to the inlet as the mass-flow-rate ratio decreases. Therefore, the minimum pressure for "State B" decreases with decreasing mass-flow-rate ratio. At low mass-flowrate ratios, the minimum pressure for "State C" decreases gradually with decreasing the mass-flow-rate ratio that reflects the convex downward behavior of the working fluid temperature compared with the cold water temperature. The slope of the cold water temperature is steep when the cold water temperature difference between the condenser inlet and outlet increases. The range of "State B" increases with increasing the cold water temperature difference because the pinch-point for "State B" is transferred from the cold water condenser outlet to inlet.

Hence, the GMTD method can be applied to cycle calculations of power plants and refrigeration systems using ammonia/water mixture. Further research into compatibility conditions and accuracy of the GMTD method will be carried out.

REFERENCES

1. A. H. A. William and W. Chih, "Renewable Energy from the Ocean: A Guide to OTEC," Oxford University Press, New York, 1994.
2. Y. Ikegami and H. Uehara, "Conceptual Design of Ocean Thermal Energy Conversion (OTEC) Power Plants in Sri Lanka," Proceedings of the 8th International Society of Offshore and Polar Engineers, 1998, pp. 140-148.
3. Y. Ikegami and H. Uehara, "Optimum Design Point for a Closed-Cycle OTEC System," Proceedings of the 4th International Society of Offshore and Polar Engineers, 1994, pp. 383-389.
4. F. Sun, Y. Ikegami, B. Jia and H. Arima, "Optimization Design and Exergy Analysis of Organic Rankine Cycle in Ocean Thermal Energy Conversion," Applied Ocean Research, Vol. 35, 1994, pp. 383-389.
5. N. Yamada, A. Hoshi and Y. Ikegami, "Performance Simulation of Solar-Boosted Ocean Thermal Energy Conversion Plant," Renewable Energy, Vol. 34, No. 7, 2009, pp. 1752-1758.
6. F. Sun, Y. Ikegami and B. Jia, "A Serformance Simulation of Solar-Boosted Ocean Thermal Energy Conversion Plant," Renewable Energy, Vol. 41, 2012, pp. 220-219.
7. A. I. Kalina, "Generation of Energy by Means of a Working Fluid, and Regeneration of a Working Fluid," United States Patent 4346561, 1982.
8. H. Uehara and Y. Ikegami, "Parametric Performance Analysis of Otec Using Kalina Cycle," ASME, 1993, pp. 203-207.
9. H. Uehara, Y. Ikegami and T. Nishida, "Performance Analysis of OTEC System Using a Cycle with Absorption and Extraction Processes," Transactions of the Japan Society of Mechanical Engineers, Series B, 64-624, 1998, pp. 384-389.
10. G. Wall, C. C. Chuang and M. Ishida, "Exergy Study of the Kalina Cycle," American Society of Mechanical Engineers (ASME), 10-15, 1989.

11. Y. M. El-Sayed and M. A. Tribud, "Theoretical Comparison of the Rankine and Kalina cycles," ASME Publication, 1, 1985, pp. 97-102.

12. V. P. Pall and L. Eliasson, "Factors Influencing the Economics of the Kalina Power Cycle and Situations of Superior Performance," International Geothermal Conference, 2003, pp. 32-40.

13. M. D. Mirolli, "Cementing Kalina Cycle Effectiveness, the Kalina Cycle for Cement Kiln Waste-Heat-Recovery Power Plants," IEEE Industry Applications Magazine, Vol. 12, No. 4, 2006, pp. 60-64.

14. U. R. Nasruddin, M. Rifaldi and A. Noor, "Energy and Exergy Analysis of Kalina Cycle System (KCS) 34 with Mass Fraction Ammonia-Water Mixture Variation," Mechnical Science and Technology, Vol. 23, No. 7, 2009, pp. 1871-1876.

15. Z. Guzovic, D. Loncar and N. Ferdelji, "Possibilities of Electricity Generation in the Republic of Croatia by Means of Geothermal Energy," Energy, Vol. 35, No. 8, 2010, pp. 3429-3440.

16. K. E. Starling, D. W. Johnson, H. Hafezzadeh, L. W. Fish, H. H. West and K. Iqbal, "Use of Mixtures as Working Fluids in Ocean Thermal Energy Conversion Cycles— Phase I," US DOE Report, 1974.

17. K. E. Starling, D. W. Johnson, H. Hafezzadeh, L. W. Fish, H. H. West and K. Iqbal, "Use of Mixtures as Working Fluids in Ocean Thermal Energy Conversion Cycles— Phase II," US DOE Report, 1978.

18. C. B. Panchal, D. L. Hillis, J. J. Lorenz and D. T. Yung, "OTEC Performance Tests of the Trane Plate-Fin Heat Exchanger," Ocean Thermal Energy Conversion Program Argonne National Laboratory, 1981.

19. Y. Ikegami, H. Goto, T. Morisaki and T. Furukawa, "Effect of Working Fluid Flow Rate and Ammonia Concentration on OTEC Using Ammonia/Water Mixture as Working Fluid," 13th Asian Congress of Fluid Mechanics, 2010, pp. 1026-1029.

20. Y. Ikegami, K. Urata, J. Inadomi, H. Goto, T. Morisaki, K. Inoue and S. Goto, "Investigation on the Stability of OTEC System Using Ammonia/Water Mixture as Working Fluid by Continuous Operation for Two Weeks," Proceedings of the 20th International Society of Offshore and Polar Engineers, 2010.

21. M. Utamura, K. Nikitin and Y. Kato, "Generalization of Logarithmic Mean Temperature Difference Method for Heat Exchanger Performance Analysis," Thermal Science & Engineering, Vol. 15, No. 3, 2007, pp. 163-172.

22. PROPATH Group, "A Program Package of ThermoPhysical Properties of Fluids, Version 11.1," PROPATH Group.

12

CHAPTER

STAR CCM+ CFD SIMULATIONS OF ENHANCED HEAT TRANSFER IN HIGH-POWER DENSITY ELECTRONICS USING FORCED AIR HEAT EXCHANGER AND PUMPED FLUID LOOP COLD PLATE FABRICATED FROM HIGH THERMAL CONDUCTIVITY MATERIALS

Kevin R. Anderson[1], Matthew Devost[1], Watit Pakdee[2], Niveditha Krishnamoorthy[3]

[1]Nonlinear FEA/CFD Multi-Physics Simulation Laboratory, Department of Mechanical Engineering, California State Polytechnic University, Pomona, USA
[2]Department of Mechanical Engineering, Faculty of Engineering, Thammasat University (Rangsit Campus), Bangkok, Thailand
[3]CD-Adpaco, One Technology Drive, Irvine, CA, USA

ABSTRACT

As telecommunication and RF power electronics applications continue to push the envelope of waste heat dissipation, more and more, we see a need for active thermal control employing forced air electronic cooling fans in unison with pumped fluid loops in order to meet temperature and performance requirements. This research paper presents results of applying Computational Fluid Dynamics (CFD) commercial industry STAR-CCM+ software for heat transfer and fluid flow simulation of a novel heat exchanger/cold plate fabricated from k-core high thermal conductivity material in order to realize thermal control system hardware design for very much applications to very large power density (~1 kW/m^2) electronics packaging scenarios. Trade studies involving different heat exchanger/cold plate materials, as well as various fault scenarios within a mock-up of a typical electronics system, are used to illustrate the upper bounds placed on the convective heat transfer coefficient. Agreement between our present findings and previous research in the field of electronics cooling is presented herein.

Keywords: CFD; High Thermal Conductivity; Heat Sink; Cooling Plate; Thermal Control

1. INTRODUCTION

The novelty of the current research focuses on a system's level approach to using CFD to examine "cause-and-effect" scenarios in an electronics cooling application. In particular, our thermal control solution uses a very high thermal conductivity material known as "k-core" [1] in order to implement a heat exchanger/cold plate used to cool high power density electronics packaging. Effective heat spreading has been implemented by previous investigators interested in heat transfer augmentation including the work of [2], which used anisotropic heat plate with preferential thermal conductivity to augment thermal performance. In the study of [3], the critical heat flux of a heat spreader was quantified. While in [4], the use of diamond coated materials was researched. This current paper utilizes Computational Fluid Dynamics (CFD) in order to study various causes and effects at a system's level. The CFD work related to electronics currently available in the literature can be broken into the following categories based on level of characteristic thermal/fluid length scale, or overall size of the hardware being analyzed: MEMS level; chip/component level; Printed Circuit Board (PCB) board level; hand-held devices (cell phones, instrumentation, etc.); desk-top computer level; intermediate compact packaging level; rack-mounted server level; very large length scale (i.e. power grid distribution applications, etc.). Herein we are interested in the intermediate compact packaging level.

At the chip/component/PCB level, various researchers have used CFD to study electronics cooling with the use of a heat sink including the work of [5], which implements CFD to characterize the performance of carbon composite and copper heat sinks on the heat transfer in a computer CPU. The work of [6] utilizes CFD to perform trade studies on the modeling of finned heat sinks, while [7] has married experimental and CFD analysis to quantify the effects of heat sinks and base plates on heat transfer behavior. Finally the work of [8] performs CFD and presents heat transfer correlations for the optimization of fin heat sinks. At a desk-top computer level, CFD has been used to study the heat transfer in computer chassis and computer box such as the work in [9] which focuses on computer chassis heat transfer performance. The study of [10] implements CFD to investigate PCB heat transfer in a computer. The study of [11] investigates computer heat transfer and cooling using a heat-pipe. More generic analyses of using CFD to electronic systems are given in the work of [12-14], wherein system's level analysis is performed using CFD to gain insight on heat transfer and thermal performance of various electronic devices, ranging from hand-held devices to desktop computers. The tutorial offered by [15] allows the novice to gain insight on how to use electronics packaging design lay-out strategies to realize more effective heat transfer flow field patterns. In the arena of rack mounted servers, several pertinent CFD based investigations have been carried out over the last decade or so. The most noteworthy of these is the study of [16], wherein a CFD tool is implemented to study the temperature distribution in rack mounted servers. The investigation of [17] uses a liquid cooling thermal management system to enhance the heat transfer in a rack cooling data center application, while the work of [18] outlines a road-map for future CFD work related to rack mounted cooling systems. For the intermediate length scale, the works of [19,20] address this type of hardware, but the studies only address moderate power density heat transfer thermal control strategies. The work of [19] includes the fidelity of the thermal vias, solder, PCB, and shelf interface. Other applications of using numerical heat transfer and CFD in electronic and avionic systems are given in [21,22]. Presently, there appears to be a void of CFD simulation databases addressing the so-called intermediate application level. The goal of this current study is to address this absence of literature by simulation of an intermediate sized electronics enclosure, which is subject to very large power densities. This is in an effort to address the ever changing requirements of affording more power in a smaller packaging level.

To date, CFD of intermediate level packaging highpower density enclosures has not been carried out comprehensively. Furthermore, there have been few or no CFD simulations of electronics cooling using CD-Adapco's STAR-CCM+ CFD software in the literature, mainly ICEPAK and FLOTHERM, which have the tools of choice as evidenced in the works of [5,6,9,12-15]. To address this absence in the literature, the current CFD simulations focus on using STAR-CCM+ CFD to simulate heat transfer and fluid dynamic behavior in intermediate sized Telecommunication/Observatory Science/Electronics/RF/Avionics systems fabricated primarily from COTS (Commercially Available Off The Shelf) components (i.e. fans, heat-sinks, cold plates, etc.) to implement the thermal control solution. This current study focuses on cost-effective thermal control system alternative, in contrast to more cost-prohibitive solutions such as those involving heat pipes, nano-fluids and/or phase change materials. The current study uses CFD to quantify the effects of heat transfer augmentation using high thermal conductivity materials in high power density electronics cooling scenarios.

2. HEAT SPREADER THERMAL TECHNOLOGY

Herein, "k-core" thermal conduction plates available from Thermacore, Inc. [1] are used in conjunction with a pumped fluid loop to illustrate the effects of heat transfer spreading. K-core heat spreader cold plates afford highefficiency heat transfer in the absence of moving parts, and are invaluable in applications where space, volume, seamless hardware integration and weight constraints dictate thermal design solutions. Physically, k-core base heat spreader cold plates transfer heat from sources such as high power dissipating electronic components, rejecting it to ambient air (through forced or natural convection) or liquid coolants (water, glycol/water mixtures, dielectric fluids, polyalphaolefin or PAO). These types of k-core cold plates are well suited for operating in harsh environments, including extreme temperatures/humidity, shock and vibration. For cooling high-power density applications, such as semiconductors, lasers, power generation, medical equipment, transportation, military electronics, etc., and other demanding applications where air cooling is insufficient, liquid cooling (pumped fluid loop) is essential. In this case Thermacore k-core based liquid cooled cold plates offer a feasible thermal control system solution. Thermacore's custom liquid cooling component cold plates include tube-in-plate, aluminum vacumnbrazed and copper brazed types for various applications. Tube-in-plate cold plate materials consist of copper or stainless steel tubes pressed into a channeled aluminum or copper extrusion or machined plate. Thermacore also offers copper cold plates that incorporate vertical fin (micro-channel) technology or, for higher performance, our specialized powdered metal construction. Thermacore's custom liquid-cooled assemblies are designed specifically for each application's unique thermal/mechanical requirements. The physical concept of Thermacore's k-core heat transfer technology is illustrated in Figure 1.

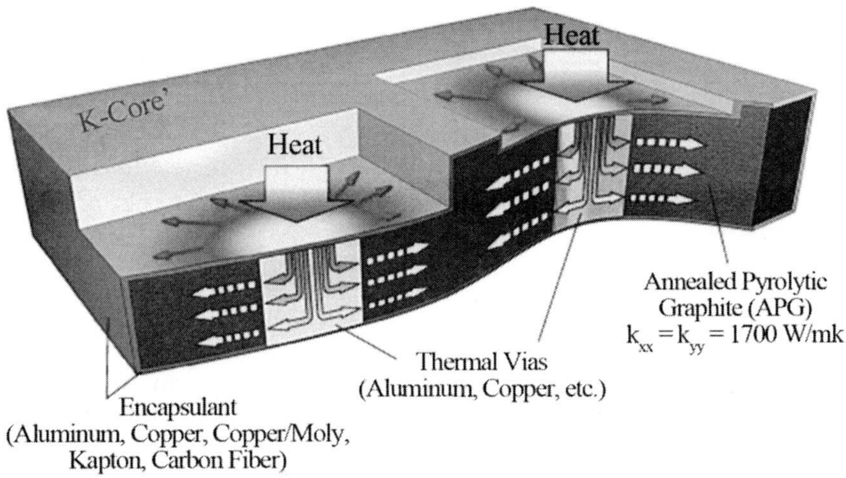

Figure 1. Thermacore k-core technology [1].

Figure 1 shows the various thermal technologies offered by Thermacore, Inc. Using Encapsulated Annealed Pryrolytic Graphite (APG) allows one to manufacture a device such as a heat spreader with heat sink integrated on one face while a labyrinth of tubing for a pumped fluid loop can be fabricated on the opposing face.

The encapsulant of Figure 1 sets the Coefficient of Thermal Expansion (CTE) and other structural properties. The encapsulant material is selected to satisfy requirements such as ease of integration into the overall system. Since the APG is anisotropic in thermal conductivity, i.e. the x-component, and y-component of thermal conductivity is very large with respect to the y-component of the thermal conductivity, the APG acts as an excellent mechanism with which to realize heat transfer enhancement via preferential heat spreading. Figure 2 illustrates the temperature dependency of the thermal conductivity of APG.

From Figure 2 it can be seen the APG has its largest thermal conductivity at 150 K (−123 C) where k= 2800 W/m-K making is a valuable thermal heat transfer material for spacecraft applications. For electronics cooling applications where the nominal temperature is on the range of 350 to 400 K (77 C to 127 C) the APG has a thermal conductivity of k = 1500 W/m-K (as compared to Copper have k = 400 W/m-K, and Aluminum = 200 W/m-K). In this current study we investigate the use of APG k-core technology for temperatures in the range of 266 K to 373 C (29 C to 100 C), k = 1600 W/m-K to k = 1100 W/m-K.

3. SYSTEM HARDWARE DESCRIPTION

Figure 3 shows the layout of the electronics being investigated in this paper. The configuration of Figure 4 was selected in order to mimic an intermediately sized high power density power electronics application. The overall system consists of a very large housing with eight identical sub-modules arranged in order to minimize overall system volume. The sub-modules were each populated with various components, totaling twelve each per submodule. These components were left to be very generic in order to facilitate our trade study involving simulation of system level electronics using CFD.

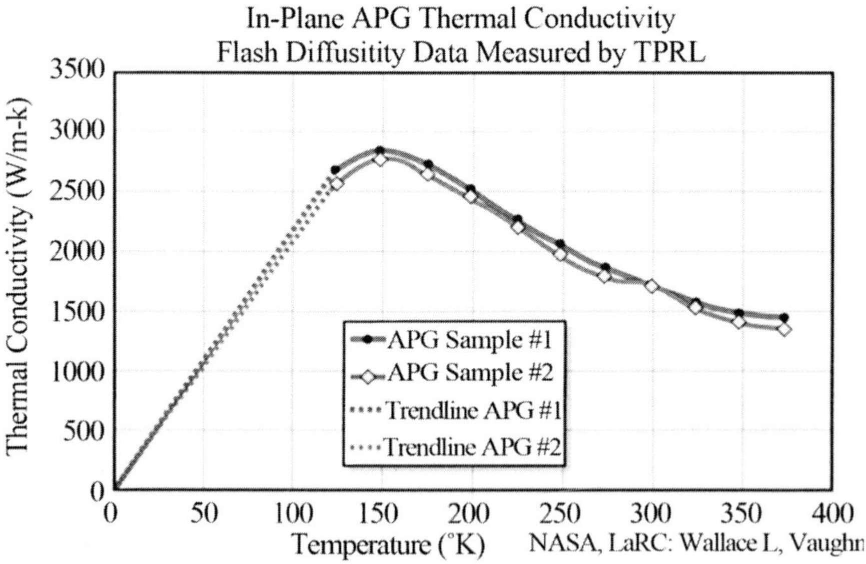

Figure 2. Thermal conductivity for APG material [1].

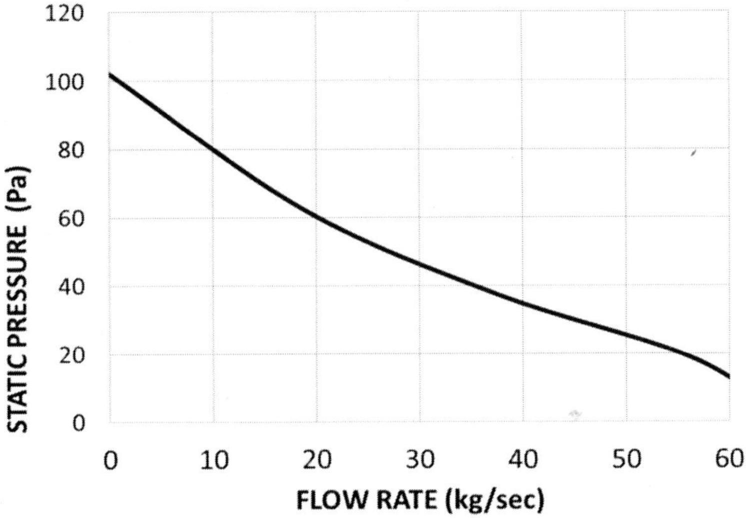

Figure 3. Fan curve specifications for thermal control hardware simulations.

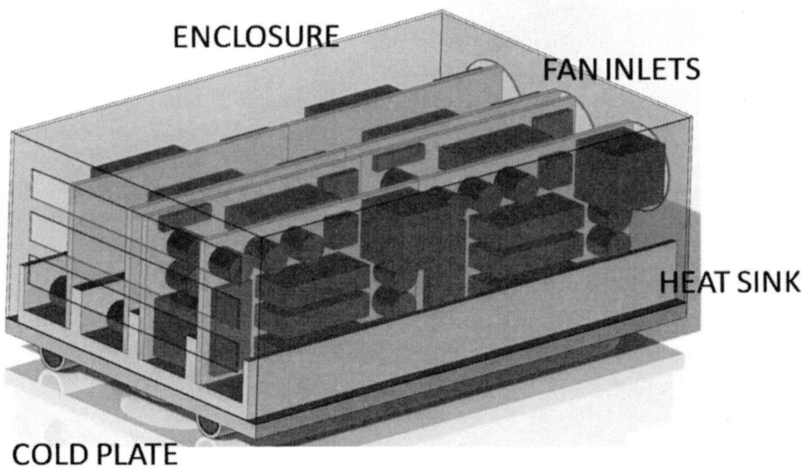

Figure 4. Electronics System Components.

As seen in Figure 4, the thermal control system is comprised of three major components; the heat sink, the cold plate and the forced convection fans/housing enclosure. The majority of heat transfer removal capacity is accomplished via the forced convection air flow of the fans, while the cold plate serves to hold a constant boundary temperature in order to act as a heat sink. This enclosure measures overall dimensions W*L*H = 0.5 m by 1.0 m by 0.5 m (19.7 inches by 39.4 inches by 19.7 inches, thus it is roughly the size of a foot-locker, smaller than a typical rack-mounted server, and larger than a typical PC workstation tower. As is traditionally done in practice in electronics manufacturing, no particular attention was paid to the exact layout of the overall system, i.e. the eight sub-modules were placed into the overall enclosure in a staggered configuration. This of course is not optimum, since some components on some of the submodules will be exposed to fans while others will not. The effects of material selection for the manufacturing of the heat exchanger and fluid loop cold plate has profound ramifications on the optimization of the heat transfer enhancement of the electronics package as a whole. Thus, the current study focuses on the effects of using the previously motivated APG k-core high thermal conductivity material in the fabrication of an integrated heat exchanger, cold plate on the convection heat transfer coefficient (aka h-value) in electronics cooling applications.

The CFD simulations presented herein utilized fully 3-d conjugate heat transfer; forced external air convection; forced internal cooling loop; solid conduction; surface to surface gray body radiation, turbulent. To mimic typical COTS cooling fans, the fan-curve of Figure 3 below was used in the simulations. Figure 3 gives fancurve characteristics for a fan rated at 30 kg/sec with 50 Pa static pressure (60 cfm with 0.2 inches W.G.).

The parametric trade space investigated over the course of our investigation is shown below in Table 1.

The electronics power listed as 500 to 1000 W in Table 1 was accomplished as per the following example, for a 720 W system, each component was assign 7.5 W, thus 7.5 W/component*12 components/sub-module*8 sub-modules/housing*1 housing = 720 W. Using the footprint of 1 m*0.5 m of the cold plate, this affords and electronics power density of 1440 W/m^2. According to Figure 5, the viable technologies for a tolerated temperature delta of 20 C involves direct air, forced convection, while for temperature deltas on the order of 100 C, a hybrid of forced air and natural convection in addition to radiation is recommended. However, in spacecraft and terrestrial applications of high power density heat transfer there is no gravity, thus natural convection cannot be relied on. Additionally, radiation effects in the presence of forced convection have been found to be minimal by prior CFD investigations [9,12,14].

Table 1. Parametric trade space for k-core thermal control system study.

Parameter	Range
Electronics Power (W)	500 - 1000
Electronics Power Density (Kw/m^2)	1 - 2
Inlet Air Flow (m/s)	5 - 15
Inlet Air Temperature (°C)	10 - 20
Water Inlet Temperature (°C)	10 - 20
Water Inlet Flow Rate (kg/s)	0.10 - 0.20
Heat Sink Thermal Conductivity (W/m-k)	200 - 1100
Cold Plate Thermal Conductivity (W/m-k)	200 - 1100

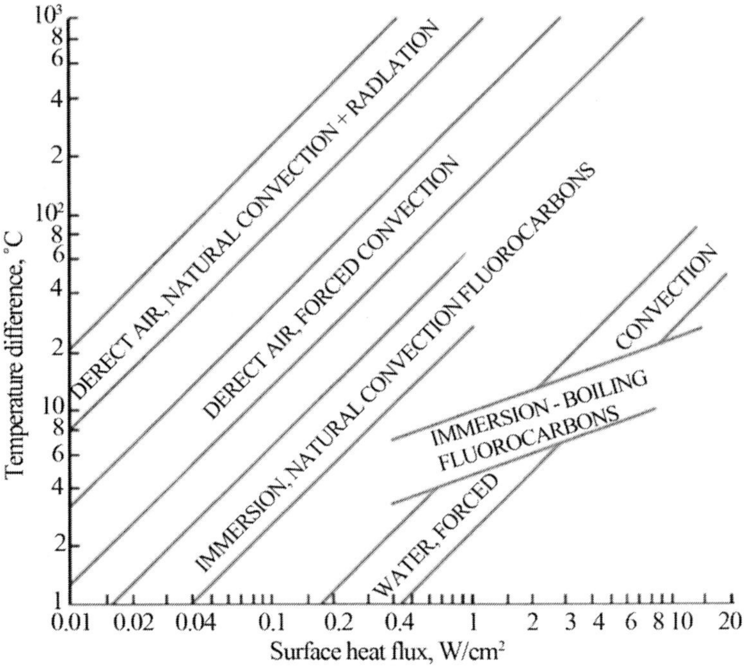

Figure 5. Roadmap of thermal technologies [23].

Thus a hybrid solution of forced convection and highly thermally conductive heat sink/pumped fluid looped thermal control system architecture is proposed herein to address spacecraft and terrestrial and avionics applications which continue to push the envelope of power size to area ratio to the extreme limits of today's capable "stateof-the-art" technologies.

4. CFD MODEL DESCRIPTION

The CFD package STAR-CCM+ from CD-Adapco was used to import the system geometry of Figure 3. The workflow in STAR-CCM CFD modeling is mush the same as in other packages, such as ANSYS-FLUEN, Flotherm, NX-Space Systems Thermal, COMSOL, etc. First the geometry is imported, then materials are assigned to components, boundary conditions are assigned, mesh is generated, solvers/models are selected, the equations of motion are then solved and results are post-processed. For the simulations outlined herein the STAR-CCM+ CFD software uses Polyhedral Unstructured Mesh with Prismatic Boundary Layer enhancement. The mesh used in the CFD simulations presented herein is shown below in Figure 6.

The software package STAR-CCM+ allows the user to specify the regions (parts) as well as the thermal interface between each part. The cooling fans are modeled in STAR-CCM using an in-place interface whereby in Figure 5 this is illustrated with the dark blue circles situated at the northwest portion of the diagram. These in-place interface fans allow the used to assign a fan-curve of the form of that given in Figure 5 to be keyed in as a polynomial curve fit. STAR-CCM+ also allows the user to model the swirl of the fan, however for this system's level study, this option was found to be too computationally expensive, since we have three fans at the inlet of the housing to simulate. The impact of this modeling assumption if felt to be minimal, since our primary objecttive herein was to simulate the flow of air through the entire enclosure with a fan rated at a certain volumetric flow rate and static pressure drop, the use of interface fans is warranted. The study of swirl around the vicinity of the fans is not within the current scope of investigation.

4.1. Grid Independence Study

Typical cell count for the system level CFD model was on the order of 2.5 million cells. Five prismatic boundary layer cells were employed to resolve the boundary layer at all solid-fluid interfaces. Various trades were run in order to determine the optimum mesh size, for 2.5 million polyhedral cells, a base element size of 0.01 m was used, mesh independence studies centered around this value were performed, i.e. the base element size was varied from 0.005 m (more cells, highest fidelity mesh) to 0.02 m (fewer cells, coarsest mesh). A monitor point was selected in the model, and the primitive variables were monitored as the cell size was halved, resulting in the choice of 0.01 m (2.5 million cells) as the baseline model size. CPU run-time on a 64-bit quad-core DELL PC workstation was on the order of six hours.

4.2. Governing Equations

Headings, The governing equations solved in STARCCM+ via the Finite Volume Method for our laminar, conjugate heat transfer simulations are the incompressible Navier-Stokes equations stated below in continuous integral form

$$\frac{\partial}{\partial t}\int_V \boldsymbol{W}\chi \mathrm{d}V + \oint [\boldsymbol{F}-\boldsymbol{G}]\cdot \mathrm{d}\boldsymbol{a} = \int_V \boldsymbol{H}\mathrm{d}V \tag{1}$$

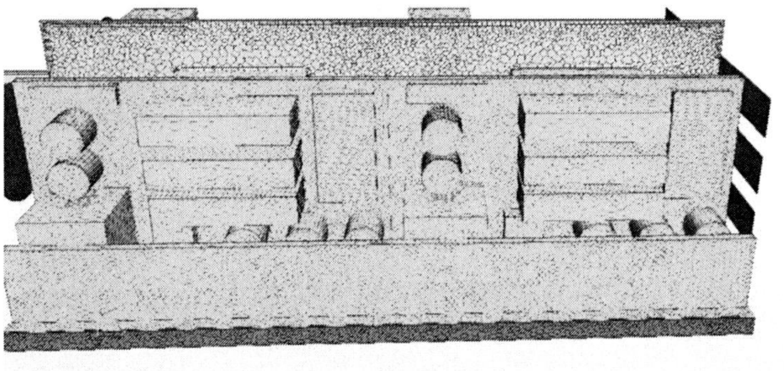

Figure 6. CFD mesh used in simulations.

$$\boldsymbol{W} = \begin{Bmatrix} \rho \\ \rho \boldsymbol{u} \\ \rho E \end{Bmatrix}, \boldsymbol{F} = \begin{Bmatrix} \rho(\boldsymbol{u}-\boldsymbol{u}_g) \\ \rho(\boldsymbol{u}-\boldsymbol{u}_g)\otimes \boldsymbol{u} + p\tilde{I} \\ \rho(\boldsymbol{u}-\boldsymbol{u}_g)H + p\boldsymbol{u}_g \end{Bmatrix},$$

$$\boldsymbol{G} = \begin{Bmatrix} 0 \\ \tilde{T} \\ \tilde{T}\cdot \boldsymbol{u} + \dot{\boldsymbol{q}}'' \end{Bmatrix}, \boldsymbol{H} = \begin{Bmatrix} S_u \\ \boldsymbol{f} \\ S_u \end{Bmatrix} \tag{2}$$

where r, $\boldsymbol{u}$, E, and p denote the density, velocity, total energy per unit mass, and pressure of the fluid, respecttively. The parameter $\tilde{T}$ is the viscous stress tensor, while $\dot{\boldsymbol{q}}''$ denotes the heat flux vector and $\boldsymbol{u}_g$ is the grid velocity vector. The total energy is related to the total enthalpy via the caloric equation of state

$$E = H - \frac{p}{\rho} \tag{3}$$

where the total enthalpy is the sum of the internal energy and the kinetic energy via

$$H = C_p T + \frac{u^2}{2} \tag{4}$$

For the remaining terms in the Navier-Stokes equation, $\boldsymbol{H}$ denotes the body force vector, $\boldsymbol{W}$ is the vector of conserved quantities $\boldsymbol{F}$ is the vector of inviscid terms, $\boldsymbol{G}$ is the vector of viscous terms, $\tilde{I}$ is the identity tensor, c denotes the porosity, and m is the dynamic viscosity. The constitutive equation modeling heat conduction in the solid bodies is taken to by Fourier's Law stated below in differential form

$$\boldsymbol{q}'' = -k\nabla T \tag{5}$$

The above equations are implemented on an unstructured polyhedral mesh and the resulting algrebraic equations are solved using the algebraic multi-grid method.

5. RESULTS

The results for temperature distribution, flow vectors and streamlines in our simulations are illustrated in Figures 7-9. The parameters for the simulation were as follows: each component has 7.5 W giving 720 W for 12 components and 8 sub-modules. Additionally, each PCB card had an additional 10 W of power applied to it, thus 80 W for the total package of eight cards. Hence, this simulation is 800 W over a 1 m × 0.5 m footprint giving 1600 W/m² = 1.6 kW/m², which is in the category of a very high power density application. Figure 7 shows the inlet air coming from the fans at 303 K (30°C) and a heat sink/cooling plate temperature of 283 K (10°C). Figure 8 shows the velocity vectors for the situation of Figure 7, while Figure 9 illustrates streamlines of the flow field for the system of Figure 7.

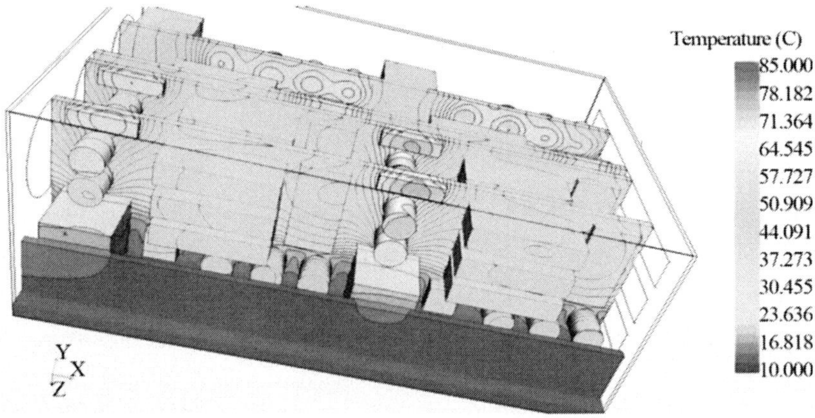

Figure 7. Component temperatures.

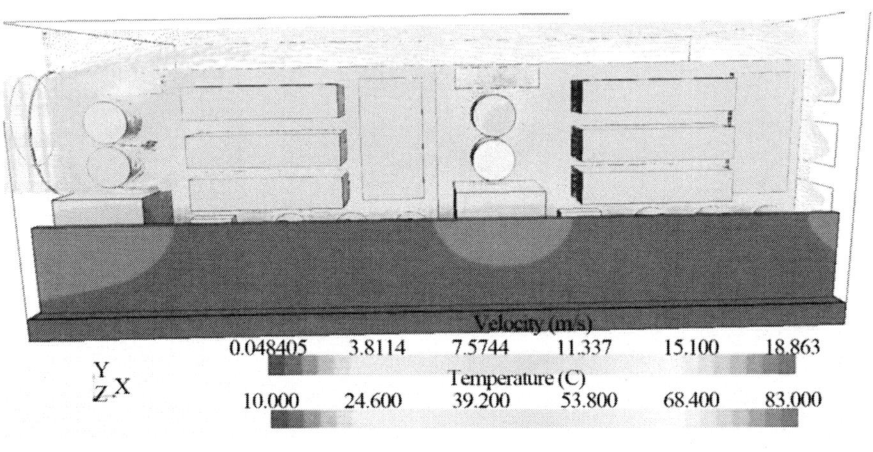

Figure 8. Velocity vectors colored by air speed and electronics components colored by temperature.

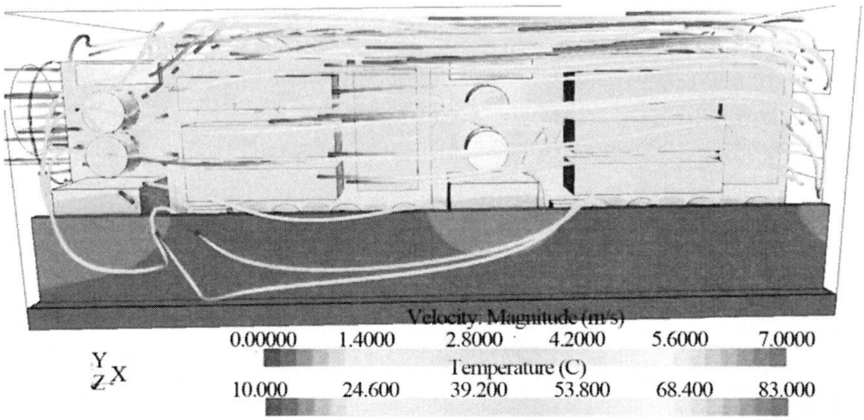

Figure 9. Streamlines colored by air stream speed and electronics components colored by temperature.

The heat sink/cooling plate assemblage serve to keep the baseplate of the system at 10°C, serving as an effective heat sink. The isothermal contours of Figures 7-9 indicate that the components witness a maximum temperature of 356.4 K (83.4°C), thus there is temperature difference of 73.4°C in this system. Next, a series of parametric studies was completed each of which is discussed in detail below.

5.1. Effect of Thermal Conductivity on Heat Transfer

The primary objective of this current research is to demonstrate the heat spreading characteristics of APG k-core material. Thus, the CFD model described above was exercised by changing the thermal conductivity of the heat sink/cold plate from a very low (bare Aluminum) to a very high value (APG k-core). The results of this parametric study are shown below in Figure 10 which shows the system temperature delta, DT (°C) plotted as a function of the heat sink/cold plate thermal conductivity k (W/m-K).

By inspecting Figure 10 the spreading effects of APG k-core become immediately obvious. The relationship between thermal conductivity and maximum system temperature delta is expected to be linear via the relationship for heat conduction given by the first law of thermodynamics

$$Q = \frac{1}{R_{\text{thermal}}} \Delta T \tag{6}$$

where the thermal resistance for conduction and contact interface is given by

$$R_{\text{thermal}} = R_{\text{contact}} + R_{\text{conduction}} = \frac{1}{h_{\text{contact}}} + \frac{x}{kA} \quad (7)$$

The terms are written in series to indicate that the heat must flow from the heat sink via conduction to the cold plate with an intermediate resistance due to contact in between them. For this study we assumed that the contact h-value for the contact conductance was very large, i.e. since the heat sink/cold plate will ultimately be fabricated as one cohesive unit.

5.2. Heat Transfer Coefficient Comparison

One very important piece of information valuable to the electronics cooling community continues to be the elusive film convection heat transfer coefficients. To this end, Figure 11 shows values of convective heat transfer coefficient (W/m²-K) for a typical CFD simulation of our study. From Figure 11, the convective heat transfer values are found to range from $20 < h < 70$ W/m²-K for the sub-modules, PCB cards, and component assembly. These values are in agreement with the heat transfer literature [24] for systems of this category.

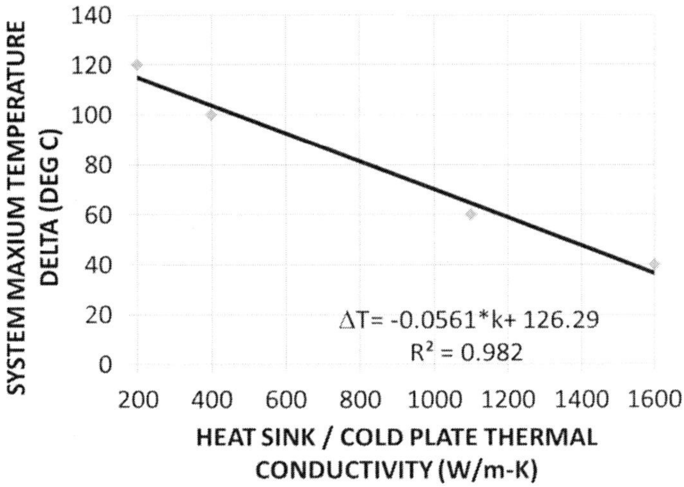

Figure 10. Thermal conductivity effect on heat spreading in CFD simulations.

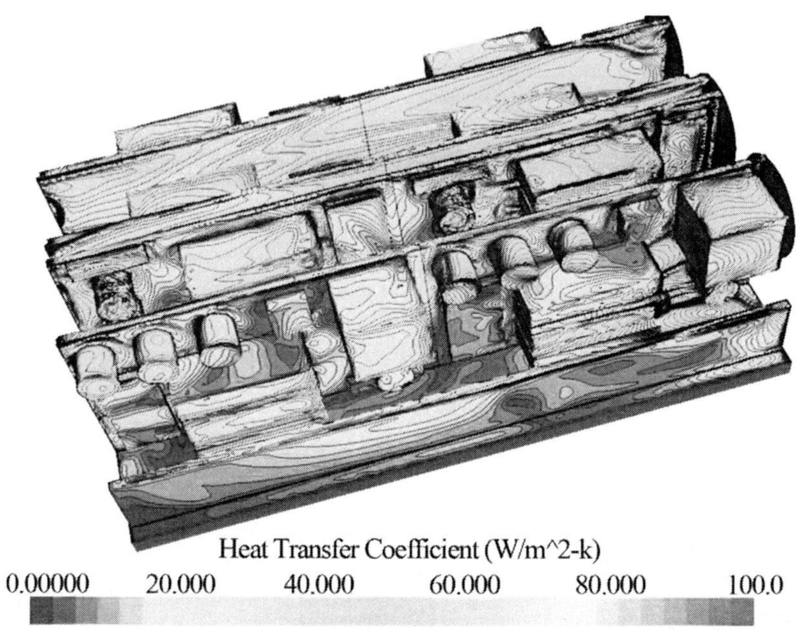

Figure 11. Convective heat transfer coefficient contours.

In order to quantify the effect of inlet fan air speed upon the heat transfer coefficient behavior, a series of CFD simulation were performed whereby the flow rate of air was varied, thus resulting in a different average velocity emanating from the fan. Figure 12 shows the results of our CFD simulations in comparison with the laminar flat plate Nusselt number based correlation of [25]. The correlation of [25] is based on the Nusselt number for flat plate laminar flow, and is written as

$$h = 0.00109\sqrt{\frac{V}{L}} \tag{8}$$

with V (ft/min) and L (in). As expected, the handbook correlation of [25] is conservative, i.e. it over predicts the heat transfer coefficient for the range of flow speeds investigated. This is expected, as the uncertainty in a standard textbook based correlation is at least 25% and that the handbook based correlations should not be viewed as sacrosanct [24]. The real mediation to resolve this uncertainty is to perform a system's level test on actual engineering hardware, and then perform a correlation of the numerical model. The data from our CFD study are found to be in qualitative agreement with the simulations and correlations offered by the research of [11,26], which both present plots of Nusselt number versus Reynolds number (i.e. heat transfer coefficient versus flow speed) for heat sinks in electronics cooling applications.

Next, the heat transfer coefficient of the system is plotted against temperature delta of the system as shown in Figure 13. Figure 10 shows convective heat transfer coefficient h (W/m^2-K) versus temperature delta ($T-T_o$). The CFD data points are assigned a power-law curve fit and compared to the correlation offered by [23]. The recommended correlation of [23] for electronics cooling is given by

$$h = 2.44\left(\frac{\Delta T}{L}\right)^{1/4} \tag{9}$$

As shown in Figure 13, the hand-book correlation of Equation (9) is found to over-predict the heat transfer coefficient in the range of system temperature differences studied. At this point it should be reiterated that according to [24], the heat transfer correlations given by Equation (8) and Equation (9) are expected to carry an uncertainty of at least 25%. Clearly, the agreement of our current CFD results in Figures 12 and 13 are within the realm of this uncertainty range.

5.3. Effect of Cold Plate Coolant Flow Rate

Next we investigate the effects of cold plate coolant flow rate on the system thermal performance. A series of CFD simulations were carried out varying the inlet flow rate of the water in the cold plate from 0.1 kg/sec to 0.2 kg/sec (1.6 gpm to 3.2 gpm for water with r = 1000 kg/m^3). The results are shown in Figure 14 where the heat transfer coefficient of the system is plotted as a function of the flow rate in the cold plate.

It should be clarified here that the influence of convective heat transfer coefficient of the entire system, as shown in Figures 7-9, i.e. components and PC boards is plotted in Figure 14, rather than the internal heat transfer coefficient of the tubes comprising the cold plate. As can be seen in Figure 15 the heat exchanger/cold plate assembly maintains a relatively constant temperature displaying only a 5°C overall spatial gradient. This is primary advantage of utilizing a pumped fluid system cold plate apparatus, i.e. it gives one the ability to hold a somewhat isothermal boundary condition in an actual thermal system.

5.4. Effect of Contact Resistance

The next segment of our investigation deals with simulation of the thermal gasketing material comprising the contact conductance interface between the various components in the system. Typical thermal gasket materials used extensively in the commercial electronics cooling industry include Chotherm [26], Thermal Grease [27] and Grafoil [28]. These materials have advertised nominal contact heat transfer coefficients of 5000 W/m^2-K for Chotherm, 50,000 W/m^2-K for Grafoil, and 10,000 W/m^2-K. In practice it is best to apply an area void fraction of at least 50% to these nominal values in order to account for that fact that uniform pressure is not available between the two interfaces being mounted together. This puts a more conservative bounds of the values or thermal contact resistance for these materials as follows, R = 4 × 10^{-4} K-m^2/W for Chotherm, R = 4 × 10^{-5} K-m^2/W for Grafoil, and R = 2 × 10^{-4} K-m^2/W for Thermal Grease. Using these values as guidance, CFD simulations were performed on the system level model, whereby all of the interfaces were set to have a contact interference with the values of contact resistance in the range $4 \times 10^{-4} < R < 2 \times 10^{-3}$ K-m^2/W ($2500 < h < 500$ W/m^2-K) thus emulating a "very good" thermal interface to a "very poor" thermal interface. These results are shown in Figure 16 which plots maximum temperature in the system versus thermal contact resistance.

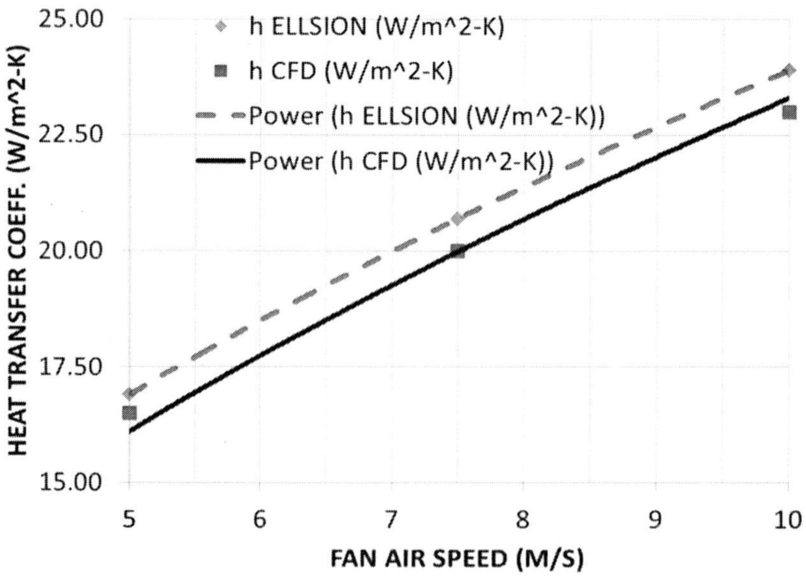

Figure 12. Heat transfer coefficient versus fan air speed.

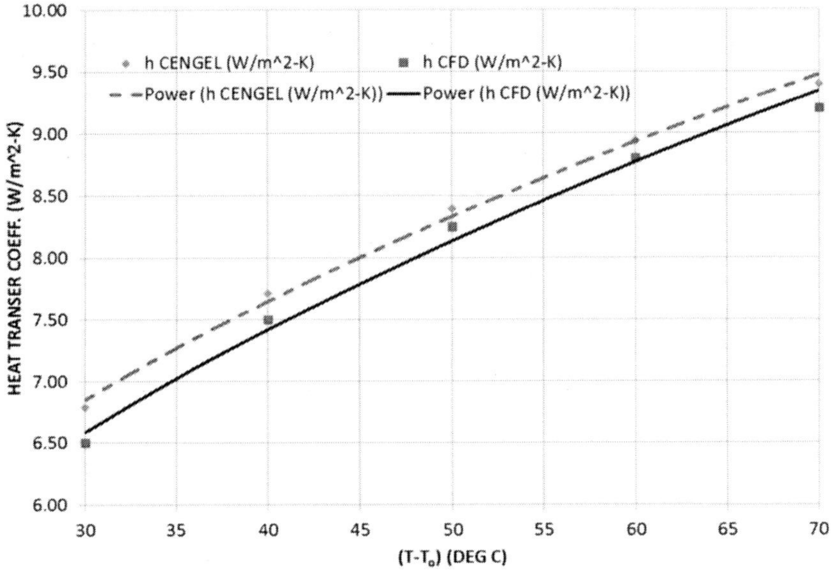

Figure 13. Heat transfer coefficient comparison.

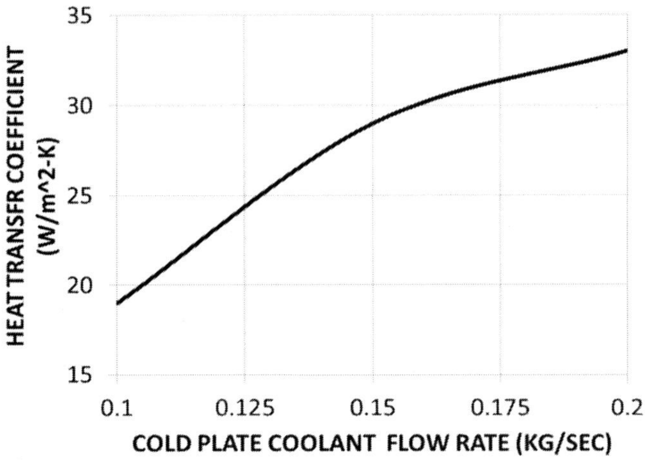

Figure 14. Heat Transfer coefficient versus cold plate flow rate.

As expected the trend of Figure 16 is linear, since the thermal contact resistance is embedded in the conduction heat transfer model of Equation (7) as previously discussed. Figure 17 shows isothermal contours for a submodule which has a very poor thermal contact interface, i.e. R = 2×10^{-3} K-m²/K.

Figure 17 shows the range of temperatures, from smallest of 286 K (13°C) where the PCB card is in contact with the heat sink/cold plate assembly, to a largest value of 350 K (77°C) where the warmest component on the PCB is at the extremity of the PCB card near the air flow region. Here, the impact of a poor thermal contact interface is profound, leading to a 64°C temperature spatial thermal gradient across this particular sub-model.

5.5. Thermal Runaway Due to Faulty Components

Thermal runaway due to faulty components in the system are readily simulated using a turn-key CFD model of the sort presented in this paper. Figure 18 shows the effect of simulation of a bad component due to an internal manufacturing flaw(s), an internal short, etc. on a particular PC board. The component experiencing thermal runaway is immediately identified as the outlier in Figure 18, with a case temperature of 580 K (307°C) while the interface of the sub-modules with the heat sink is 290 K (17°C), thus a delta of 290°C is manifested by this single component failure simulation.

Figure 15. Heat sink/cold plate temperatures.

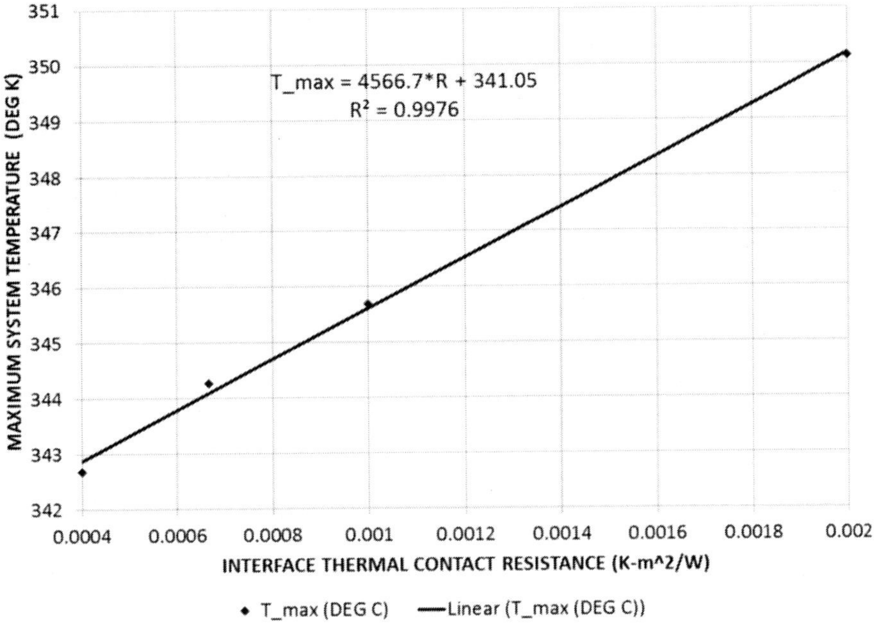

Figure 16. Thermal fault simulation showing effect of contact resistance.

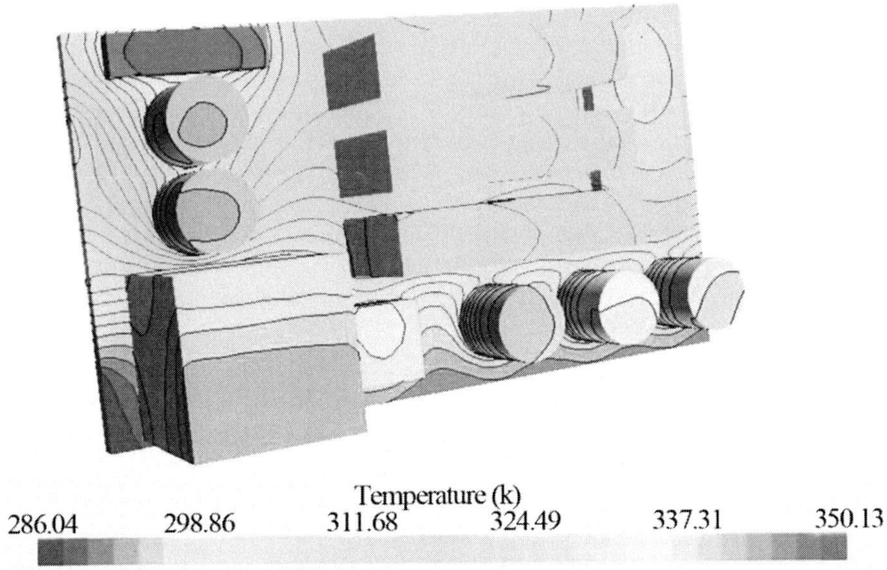

Figure 17. Effect of poor thermal contact resistance.

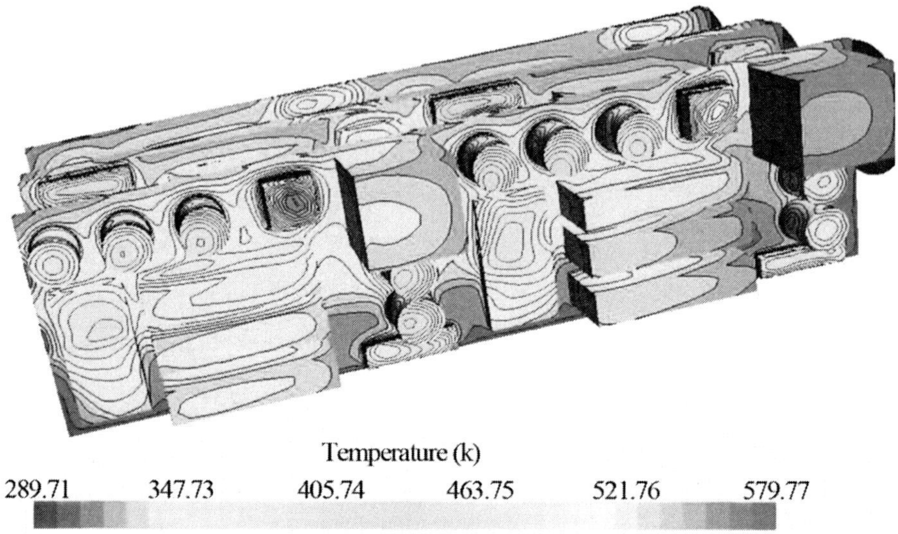

Figure 18. Thermal fault simulation of defective chip.

5.6. Fan Outage Simulations

One very common lead to failure in large data racks and thermally controlled avionics electronics systems is fan failure. This leads to the thermal designer having to carry a redundant system, i.e. back-up fans, in case the primary fan cooling system experiences failure due to power drop out and/or fan motor burn-out. In order to understand the effects of fan failure, various CFD simulations were performed in by varying the air flow around a nominal value of 6 m/s (for a 80 mm dia. fan rated at 60 cfm). Figure 19 shows the results of the fan failure parametric trade study. Figure 19 is a plot of the system level maximum component/chip temperature versus fan air speed.

The trend depicted in Figure 19 is to be expected, i.e. as the fan speed approaches zero, the various components and PC cards within the sub-modules will witness a large temperature. The results of Figure 19 are in agreement with the trends of thermal impedance $\Delta T/Q$ versus Reynolds number identified in the study of [29] which focused on the free stream cooling heat transfer cooling of micro-processor chips.

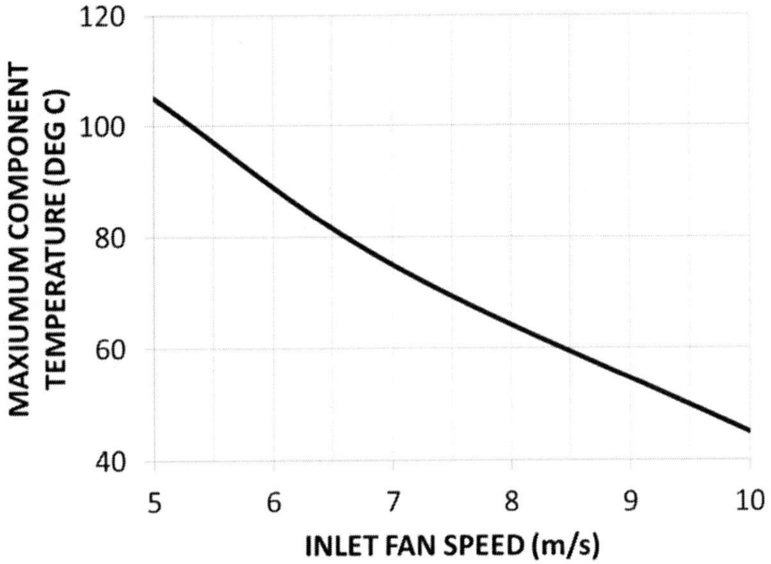

Figure 19. Maximum component/chip temperature versus cooling fan inlet air speed.

6. CONCLUSION

This paper has presented CFD simulations of heat and fluid flow behavior in a moderate sized package of electronics undergoing very large power dissipations on the order of 1000 W/m^2. A combination of heat transfer enhancement thermal control strategies (high thermal conductivity materials, forced convection cooling using DC computer fans, and internal forced convection heat transfer using a pumped fluid loop cold plate/heat exchanger system) have been combined in order to simulate the effectiveness of implementing COTS components to realize a thermal control solution. The results obtained herein are in qualitative agreement with previous studies involving electronics cooling simulations [7,9,13,16]. Parametric exercises involving air flow rate, water flow rate, thermal interface material, defective electronic components, and failed fans have been carried out.

REFERENCES

1. "Thermacore k-Core Data Sheet," 2013. http://www.thermacore.com/products/kcore.aspx
2. D. D. L. Chung and Y. Takizawa, "Performance of Isotropic and Anisotropic Heat Spreaders," Journal of Electronic Materials, Vol. 41, No. 9, 2012, pp. 2580-2587.
3. Y. S. Yoon, H. Yang and H. Y. Kwak, "Enhancement of the Critical Heat Flux by Using Heat Spreader," KSME International Journal, Vol. 17, No. 7, 2003, pp. 1063- 1072.
4. K. J. Gray, "Effective Thermal Conductivity of a Diamond Coated Heat Spreader," Diamond and Related Materials, Vol. 9, No. 2, 2000, pp. 201-204.
5. P. Mohan, and P. Govindarajan, "Thermal Analysis of CPU with CCC and Copper Base Plate Heat Sinks Using CFD," Heat Transfer Asian Research, Vol. 40, No. 3, 2011, pp. 217-232.
6. A. Part, R. Linton and D. Agonafer, "Coarse and Detailed CFD Modeling of a Finned Heat Sink," IEEE Transactions on Components, Packaging and Manufacturing Technology, Vol. 18, No. 3, 1995, pp. 517-520.
7. P. Mohan and P. Govindarajan, "Experimental and CFD Analysis of Heat Sinks with Base Plate for CPU Cooling," Journal of Mechanical Science and Technology, Vol. 25, No. 8, 2011, pp. 2003-2012.
8. J. Yang, W. Wang and L. Huazhi, "2-Dimensional CFD Simulation and Correlation Development for Optimization of Fin Heatsinks in Electronic Cooling," Journal of Thermal Science, Vol. 10, No. 4, 2001, pp. 363-371.
9. E. Ozturk and I. Tari, "CFD Modeling of Forced Cooling of Computer Chassis," Engineering Applications of Computational Fluid Mechanics, Vol. 1, No. 4, 2007, pp. 304- 313.
10. V. Eveloy, J. Lohan, and P. Rodgers, "A Benchmark Study of Computational Fluid Dynamics Predictive Accuracy for Component-Printed Circuit Board Heat Transfer," IEEE Transactions on Components and Packaging Technologies, Vol. 23, No. 3, 2000, pp. 568-578.
11. I. Tari and Y. Fidan-Seza, "CFD Analyses of a Notebook Computer Thermal Management System and a Proposed Passive Cooling

Alternative," IEEE Transactions on Components and Packaging Technologies, Vol. 33, No. 2, 2010, pp. 443-452.

12. A. Part and P. G. Tucker, "CFD Applied to Electronic Systems: A Review," IEEE Transactions on Components, Packaging, and Manufacturing Technology, Vol. 20, No. 4, 1997, pp. 518-529.

13. T. Y. Lee and M. Mahalingam, "Application of a CFD Tool for System Level Thermal Simulation," IEEE Transactions on Components, Packaging and Manufacturing Technology—Part A, Vol. 17, No. 4, 1994, pp. 564-572.

14. T. Y. Lee, B. Chambers and M. Mahalingam, "Application of a CFD Technology to Electronic Thermal Management," IEEE Transactions on Components, Packaging and Manufacturing Technology—Part B," Vol. 18, No. 3, 1995, pp. 511-520.

15. S. Aradag, U. Olgun, F. Aktuk and B. Baskbuyuk, "CFD Analysis of Cooling Electronic Equipment as an Undergraduate Project," Wiley Periodicals Inc., 2009.

16. J. Choi, Y. Kim, A. Sivasubramaniam, J. Srebric, Q. Wang and J. Lee, "A CFD Based Tool for Studying Temperature in Rack Mounted Servers, IEEE Transactions on Computers," Vol. 57, No. 8, 2008, pp. 1129-1142.

17. A. Almodi, A. Thompson, N. Kapur, J. Summers, H. Thompson G. and Hannah, "Computational Fluid Dynamics Investigation of Liquid Rack Cooling in Data Centres," Applied Energy, Vol. 89, No. 1, 2012, pp. 150-155.

18. J. Rambo and Y. Joshi, "Modeling of Data Center Airflow and Heat Transfer: State of the Art and Future Trends," Distributed Parallel Databases, Vol. 21, 2007, pp. 193-225.

19. R. Boukhanouf and A. Haddad, "A CFD Analysis of an Electronics Cooling Enclosure for Application in Telecommunication Systems," Applied Thermal Engineering, Vol. 30, No. 16, 2010, pp. 2426-2434.

20. Z. Khatir, J. Paton, H. Thompson, N. Kapur, V. Toropov, M. Lawres and D. Kirk, "Computational Fluid Dynamics (CFD) Investigation of Air Flow and Temperature Distribution in a Small Scale Bread-Baking Oven," Applied Energy, Vol. 89, No. 1, 2012, pp. 89-96.

21. D. B. Flowers and K. R. Anderson, "Numerical Simulation of Conduction Heat Transfer in a System of Slowly Rotating Concentric Shells Separated by Small Annular Gap Distances," Numerical Heat Transfer Part A, Vol. 46, No. 9, 2004, pp. 1-17.

22. K. R. Anderson, "CFD Analysis of OPALS Sealed Enclosure Electronics Sub-System," Proceedings from the Thermal & Fluids Workshop, NASA JPL, Pasadena, 2012.

23. Y. A. Cengel, "Heat Transfer—A Practical Approach," McGraw-Hill, New York, 2010.

24. F. P. Incropera and D. P. Dewitt, "Heat Transfer," McGraw-Hill, New York, 1991.

25. G. Ellison, "Thermal Computations for Electronics—Conductive, Radiative, and Convective Air Cooling," CRC Press, Boca Raton, 2011.

26. "Graftech Data Sheet," 2013. www.graftech.com

27. D. G. Gilmore, "Spacecraft Thermal Control," Aerospace Press, El Segundo, 2002.

28. "Chotherm Data Sheet," 2013. www.chomerics.com

29. M. A. Ismail, M. Z. Abdullah and M. A. Mujeebu, "A CFD Based Analysis on the Effect of Free Stream Cooling on the Performance of Micro Processor Heat Sinks," International Communications in Heat and Mass Transfer, Vol. 35, No. 6, 2008, pp. 771-778.

INDEX